The Moon's Acceleration
and Its Physical Origins

Other books by Robert R. Newton

*Ancient Astronomical Observations and the Accelerations of
    the Earth and Moon*
Medieval Chronicles and the Rotation of the Earth
Ancient Planetary Observations and the Validity of Ephemeris Time
The Crime of Claudius Ptolemy
The Moon's Acceleration and Its Physical Origins, vol. 1,
    As Deduced from Solar Eclipses

ROBERT R. NEWTON

The Moon's Acceleration and Its Physical Origins

Volume 2

As Deduced from General
Lunar Observations

The Johns Hopkins University Press
Baltimore and London

The Johns Hopkins University Press, Baltimore, Maryland 21218
The Johns Hopkins Press Ltd, London

Library of Congress Cataloging in Publication Data
(Revised for volume 2)

Newton, Robert R.
   The moon's acceleration and its physical origins.

   Includes bibliographies and indexes.
   CONTENTS: v. 1. As deduced from solar eclipses.—
v. 2. As deduced from general lunar observations.
   1. Moon—Observations.   2. Acceleration (Mechanics)—
Observations.   3. Moon—Origin.   I. Title.
QB581.N54   523.3′3   78-20529
ISBN 0-8018-2216-5 (v. 1)
ISBN 0-8018-2639-X (v. 2)

To the memory of Ralph Edward Gibson,
whose vision and leadership made this work possible.

CONTENTS

CONTENTS (Continued)

CONTENTS (Continued)

FIGURES

TABLES

Several years ago I began a program to study the secular acceleration of the moon with respect to solar time and, in particular, its dependence on time within the historic period, by using all suitable observations I could find from the present back to the earliest observations available. It soon became clear that a work of this scope could not be accommodated within a volume of reasonable size. Therefore I divided The Moon's Acceleration and Its Physical Origins into two volumes. In volume 1, I used only records of the type which say that a solar eclipse was seen at a particular place on a particular day, but without using any ancillary information such as measurements of magnitude made by trained astronomers. In the present volume 2, I will use all other observations from which we may estimate the acceleration of the moon.

There is one exception to the statement just made about volume 1. For the period since the introduction of the telescope and the pendulum clock into observational astronomy, I used the smoothed observations of the sun, moon, and planets that Spencer Jones [1939] used in his famous paper establishing the relations among the accelerations of the sun, moon, and planets.

I cite a reference by giving two items of information. The first is the name of the author underlined, as with Spencer Jones above, and the second is the date of publication (or of writing for works that were not originally published in the modern sense). The date is placed in square brackets after the name. If a work is anonymous, as many ancient and medieval works are, I use an appropriate portion of the title in place of the author's name. If a work is issued by an organization, with no authors listed, I use the name of the organization. If the year of writing is unknown, I make a reasonable estimate of the year and prefix it by the symbol "ca."

When it is necessary to specify a particular part of a reference, such as a page number or section number, I give the relevant information after the year in the square brackets. If a work is cited several times within a short space, I often omit the year after the first citation.

Dates in this work will be given in astronomical style, with the months being identified by their English names. The names of the months are written in full in the text, but only the first three letters of the English name are used in tables, in order to make the columns uniform. I believe that the style is obvious except for one point. The year before the year 1 of the common era is written as 1 B.C.E. in historical style and as 0 in astronomical style. The preceding year is written as 2 B.C.E. in historical style and as -1 in astronomical style, and so on. All records used in this work will be dated by means of the Julian calendar.

I gratefully acknowledge the support of this research by the Department of the Navy through its contract with The Johns Hopkins University, and I thank the Director of the Applied Physics Laboratory for encouraging me in this study. I particularly thank Mrs. Gayle Snyder and Mr. J. W. Howe, of The Applied Physics Laboratory, for their valuable help in preparing this work for the press.

The Moon's Acceleration
and Its Physical Origins

CHAPTER I

INTRODUCTION

1.   Ephemeris and Solar Time

Until about 1960, the ephemerides in such works as the American Ephemeris and Nautical Almanac [published annually] were based upon solar time as the independent variable.† Solar time is based upon the point in the heavens called the fictitious mean sun. When this point passes through the meridian of Greenwich, solar time is equal to 12 hours. In full, this time is called Greenwich mean solar time, but I usually write only the last two words for brevity.

Since 1960, the ephemerides have been based upon ephemeris time as the independent variable. The idea back of ephemeris time is the following: When we use solar time, we find that all the motions in the solar system have secular accelerations beyond those due to a strictly gravitational theory (specifically, the theory of general relativity). However, the accelerations of the sun and planets are proportional to their mean motions. This means that the accelerations are actually due to the time base used and that they are not due to forces of non-gravitational origin affecting the orbital motions of the earth and planets. Instead, they must be due to an acceleration of the earth's spin.

According to this hypothesis, the only motions in the solar system that have secular accelerations are the orbital motion of the moon and the spin of the earth. Both are known to be affected by the friction in the tides, and the earth's spin is also affected by several other forces. Of course, if we knew enough about the spins of other objects in the solar system, we might find that they also have secular accelerations.

In spite of the fact that ephemeris time is used in all modern ephemerides, I have used solar time as the time base in most of my work on the secular accelerations, and in particular I used it in Newton [1979]. There were two main reasons for this choice:

†Actually, the variable used was called universal time, which differs conceptually from solar time: Universal time is based upon observations of the vernal equinox while solar time is based upon observations of the sun. In practice, the difference between the two is not important. I have estimated that solar and universal times have probably disagreed by less than 1 minute in the past 2000 years, and I will use the terms interchangeably.

a. The observations used were made before 1600, and the times were based upon observations of the sun with respect to the meridian of the observer. In other words, the observations were based upon solar time, if we make the necessary correction for the longitude of the observer.

b. The ancient and medieval data that have been analyzed [Newton, 1976] do not agree with the hypothesis of ephemeris time. Further, Spencer Jones [1939] made several theoretical errors in his analysis of the modern data [Newton, 1979,† Section I.4]. When his errors are corrected, it turns out that the modern data do not agree with the hypothesis either.

I want to word the last point carefully, because it has been misunderstood in the literature. The orbital equations of motion of the solar system under the action of gravitation alone are solidly grounded upon observation and experiment, and no one has thought of any forces other than gravitation that are large enough to matter at the present level of accuracy. Thus, it is hard not to accept the idea of ephemeris time. On the other hand, the ancient, medieval, and modern data are consistent with each other, but they all conflict with the idea of ephemeris time, at least on the basis of the analyses that have been performed up to the present. This poses a dilemma.

My suspicion is that the idea is correct at the level of accuracy of current observations, and that there have been systematic errors of some sort in the handling of the data. We know, in fact, that there were systematic errors in the modern observations of Venus that Spencer Jones studied, and there may be errors in the other data that are less obvious. Still, unless we can either demonstrate the necessary forces or find errors in the analysis, I will remain suspicious both of the idea of ephemeris time and of the data that seem to contradict it.

Further, even if we accept the basic idea of ephemeris time, there have still been some serious objections to using it in a study such as this one. For one thing, the basic parameters in the theories of the sun and moon that are used [Newcomb, 1895 and Brown, 1919] were originally derived using solar time as the base, on the basis of observations that had solar time as their underlying time base, and a fundamental error was made in transforming the theories to ephemeris time. For another thing, the parameters in the theories were not derived on a consistent basis in which all interacting parameters were determined simultaneously.

However, these objections have now been overcome by a new development in orbital theories, and I will use ephemeris time as the time base for the analysis in this volume.

----

†This reference is volume 1 of the present work. I will identify it simply as volume 1 in the rest of this volume.

In the next section, I will describe the nature of the error
that was just mentioned, and in the succeeding section, I
will describe the new theoretical development that was just
mentioned.

2. Reference Epochs

In my previous work on the secular accelerations, I
have used a variety of theories of the motions of the sun,
the moon, and the planets. Specifically, the theories used
have been those of Newcomb [1895] for the sun, Brown [1919]†
for the moon, Newcomb [1895a and 1895b] for Mercury and
Venus, respectively, Clemence [1949 and 1961] for Mars, and
Eckert, Brouwer, and Clemence [1951] for the other planets.

The reference epochs used in these theories pose a
problem that can be explained by using the sun as an example. If the earth and sun were the only members of the
solar system, the apparent orbit of the sun around the earth
would be an ellipse. However, the moon and the other planets perturb this orbit slightly, and the orbit is not exactly an ellipse.

Newcomb started by calculating the deviations from an
elliptical orbit that are produced by the moon and planets.
These deviations or perturbations depend upon the orbital
parameters of the other planets, which had to be determined
by matching their theories to observation. However, the
perturbations do not need to be known with high relative
accuracy, and earlier theories had given the various parameters closely enough to make the perturbations highly
accurate.

Newcomb then subtracted the perturbations from the
observed positions and obtained a set of "observations" of
the elliptical orbit of the sun. He then found the parameters of an elliptical orbit that give the best fit to these
observations.

When one develops a dynamical theory which contains
constants that are to be derived by fitting to observations,
one should take the reference epoch of the constants to be
the midpoint of the data. In the case of Newcomb's theory
of the sun, the data used ranged from about 1700 to about
1890, but Newcomb took the reference epoch to be 1900. A
similar thing happened with all the other orbital theories
mentioned a moment ago, although the reference epoch was not
always outside the time span of the data, as it was in
Newcomb's work.

The reason for taking the reference epoch to be the
midpoint of the data is the following: In developing a
theory, one almost always fails to model the force system
completely, and in addition, one may make approximations in

---

†More accurately, I have used the modifications of this
theory made by Eckert, Jones and Clark [1954].

the development. Either effect usually gives rise to an apparent acceleration of the object in question that is not reflected in the theory. However, one minimizes the resulting errors in the theory by choosing the reference epoch to be the midpoint of the data.

In Section I.3 of volume 1, I worked out the theory of the errors in the position at the epoch, and in the mean motion, that result from taking the reference epoch to be some point other than the midpoint of the data, if the position has a second derivative that is not correctly represented by the model. The most important error is in the mean motion, since it propagates as one moves away from the reference epoch. However, if the reference epoch is chosen to be the midpoint of the data, the error in the mean motion is zero. There is still a slight error in the position at the midpoint, but it is small and does not increase with time.

There is an interesting effect that the reader can easily prove for himself. Suppose we choose a set of constants of motion by fitting to observed data, and that we make an error in the second derivative when we do so. Then let us calculate the residuals, that is, the observed values minus the values calculated from the theory. The residuals will follow a parabola, within observational error, and the correct reference epoch is the time when the slope of the parabola is zero. I will use this effect later in order to find the correct reference epoch for the theory of the earth's rotation.

3.   A New Ephemeris of the Solar System

For the first time in history, I believe, there is now an ephemeris giving the positions of the sun, moon, and planets that has been derived on a consistent basis. The basis of this ephemeris is described by Newhall, Williams, and Standish [in preparation].† The general idea is as follows:

During the past few decades, there has been a great improvement in the precision of measuring the angular coordinates of the principal celestial bodies. At the same time, it has become possible to measure the distances to some of them by using radar and laser techniques. In addition, the development of the cesium clock has provided a worldwide time base (atomic time) that is stable and highly precise. Finally, developments in computer systems and in programming now make it feasible to determine large numbers of empirical constants simultaneously by fitting to observational data.

---

†I thank T. C. Van Flandern of the U.S. Naval Observatory for telling me about this new ephemeris, and I thank Wayne Warren of the National Space Science Data Center, at NASA's Goddard Space Flight Center, for preparing a taped copy of the ephemeris for me.

From the combination of these advances, we now have an ephemeris for the sun, moon, and planets that has been derived by a numerical integration of the equations of motion of these bodies and the earth. This ephemeris is known as the DE102 ephemeris. The various constants involved in this ephemeris have been determined by fitting to the types of data mentioned, using atomic time as the time base. However, the time base used in the integration is ephemeris time, not atomic time. This poses an interesting problem.

Atomic time, as currently used, is based upon a certain transition between two states of the cesium atom, and it is independent of the phenomenon of gravitation, to extremely high accuracy. Ephemeris time, on the other hand, depends upon the assumption that the gravitational force F between two bodies of masses m_1 and m_2, respectively, is equal to[†]

$$F = Gm_1m_2/r^2,$$

in which r is the distance between the bodies and G is a constant called the gravitational constant.

Atomic time and ephemeris time have been matched in rate and epoch by careful comparisons made during, say, the past two decades. If the constants that enter into the transition energy of the cesium atom, such as the charge on the electron, are indeed independent of time, and if G and the masses of the bodies in the solar system are likewise independent of time, then atomic time and ephemeris time have been and will be equal to each other except for trifling errors that have been made in matching them.

However, the universe is apparently expanding, and some theories about the origin of gravitation require this to result in a changing value of G.[‡] If G is changing linearly with time, the result will be an acceleration of atomic time and ephemeris time with respect to each other.

For the sake of brevity and definiteness in the rest of the discussion, I will assume that atomic time flows at a uniform rate and that ephemeris time may have an acceleration with respect to it. One of the goals of present astronomical research is to determine whether this acceleration is significantly different from zero.

As I said earlier, the independent variable used in the numerical integration is ephemeris time, but the observations used to determine the constants of integration were measured with respect to atomic time. This does not matter at all if there is no acceleration of ephemeris time. Further, the effects are trivial if the reference epoch or starting epoch of the integration is the midpoint of the

---

[†]Neglecting the relativistic corrections.

[‡]The mass of the sun is known to be steadily decreasing because of its radiation of energy. However, the effect of this upon ephemeris time is trivial at the present level of accuracy.

data, which it nearly is. The midpoint of the data is about 1973.5 while the reference epoch is the Julian day number 2 440 400.5. This is the midnight that began 1969 June 28. Because the two epochs are not exactly the same, the two kinds of time do not agree exactly at the reference epoch, and the rate of flow of ephemeris time at the epoch is not exactly equal to the rate of atomic time, if ephemeris time has an acceleration. However, the direct effects of the acceleration will far outweigh the errors produced by the discrepancies in epoch and rate as soon as one is a few years away from the reference epoch.

Because of the mixed nature of the time base used in the new ephemeris, it is not correct to call it ephemeris time, as I have been doing for the sake of introducing only one new concept at a time. The time base of the new ephemeris is properly called dynamical time. However, I will often use the terms dynamical time and ephemeris time interchangeably.

The positions used to find the constants of integration were transformed to the fixed coordinate system based upon the mean equinox and equator for the epoch 1950.0,† using Newcomb's theory in order to do so. The numerical integration was then performed in this fixed system rather than in the moving system referred to the mean equinox and equator of date.

Van Flandern [in preparation] has used the integration to find new values of the mean elements of the sun and moon. He first modified the theory of Eckert, Jones, and Clark [1954] for the moon by adding a few perturbations that had been previously omitted. The most important of these is probably the addition of the quantity $\delta\Omega$ to the mean longitude of the ascending node, where

$$\delta\Omega = 0''.195 \sin (13E - 8V + 313°.9) - 0''.049T^2, \qquad (I.1)$$

in seconds of arc. In this, T is time in Julian centuries of 36525 days from the reference epoch of the integration (Julian day 2 440 400.5).‡ E denotes the mean longitude of the earth and V denotes the mean longitude of Venus. The term $0''.049T^2$ is the effect of tidal friction on the position of the node. In an earlier work [Newton, 1970, pp. 218 and 287], I tried to derive this term from ancient and medieval observations of lunar eclipses. It should be no surprise that I did not succeed.

---

†This means the beginning of the Besselian year 1950, not the calendar year. The Julian day number of this epoch is 2 433 282.419, and it is about 2 hours before the midnight that began 1950 January 1.

‡In my previous writing, I have used T to denote time in Julian centuries from noon on 1900 January 0. There will be no occasion to use this reference epoch in this volume, except perhaps in a few historical discussions, so there is no risk of confusion in using T with a different reference epoch.

After he made these modifications to the lunar theory, Van Flandern calculated the position of the moon at an interval of 2 days from 1500 to 2000 and of the sun at an interval of 8 days over the same time span. He then compared the results with those taken from the numerical integration, and found new values of the mean elements to give the best fit of the theories to the integration. Since I will not use the lunar and solar theories in this work, it is not necessary to give most of the details of the new elements. Here it should suffice to remark that positions were altered by the order of 0".1 and that rates were altered by the order of 1 "/cy.† I will give Van Flandern's expressions for the mean longitudes in Section I.8 below.

One change does need discussion, however. Newcomb's value for the constant of precession (5 025.64 "/cy) needs to be increased by 1.13, to 5 026.77 "/cy. However, when the constants of integration were fitted to the observations, the observations were precessed to 1950.0 using Newcomb's value. Hence, whenever I want to refer an ephemeris position to the mean equinox and equator of date, I first precess the position from 1950.0 to 1973.5 (the midpoint of the data) using Newcomb's value, and then I precess from 1973.5 to date using the new value.

The acceleration of the moon with respect to dynamical time was taken to be -26.21 "/cy² in the integration. This is somewhat larger in magnitude than the value -22.44 that has been used before.‡ It is slightly smaller than the estimate -28.38 ± 5.72 that I formed in volume 1 (page 26).

Van Flandern also finds that the mean obliquity ε needs to be changed by the amount $\Delta\epsilon$, where

$$\Delta\epsilon = (-0".050 \pm 0".009) + (-0".040 \pm 0".001)\underline{T}. \qquad (I.2)$$

Although this has a smaller effect on calculated positions than the error in precession between 1950.0 and 1973.5, I will include it in my calculations because it would be difficult for the reader to make this modification to my results if he should want to. The expression for the mean obliquity is now:

$$\epsilon = (23°.443 \; 237 \; 9 \pm 0°.000 \; 002 \; 5)$$

$$- (0°.013 \; 025 \; 16 \pm 0°.000 \; 000 \; 28)\underline{T} \qquad (I.3)$$

$$- 0°.000 \; 000 \; 59\underline{T}^2 + 0°.000 \; 000 \; 503\underline{T}^3.$$

---

†I use the symbol " to denote a second of arc and "cy" to denote a century when I am stating the units of some quantity, but in no other context.

‡In giving values of the accelerations, I always take the units to be seconds of arc per century per century. Most of the time I will omit the statement of the units for brevity.

Van Flandern also gives the value of ε referred to noon on 1900 January 0 and to noon on 2000 January 1.

The coordinates yielded by the integration are the rectangular coordinates referred to the mean equinox and equator of 1950.0.† Specifically, the x-axis is the mean equinox of 1950.0, the z-axis is the mean (north) pole of 1950.0, and the y-axis completes a right-handed system. To use these coordinates, I first convert them to ecliptic coordinates for 1950.0. I then precess them to the ecliptic coordinates referred to the mean equinox and ecliptic of the desired date, after first precessing to the auxiliary epoch 1973.5 for the reason that was already explained. What is done from there on depends upon the circumstances. In the most elaborate case, I convert from the mean to the true equinox, correct for aberration, and finally convert to right ascension and declination using the true obliquity of date.

In order to perform the precessions needed, I use the transformation given on page 38 of Explanatory Supplement [1977].

4.   The Reference Epoch for the Earth's Rotation

If the concept of ephemeris time is valid, the sun has no acceleration when ephemeris time is used as the base. Instead, the apparent acceleration $\nu_S{}'$ of the sun with respect to solar time really arises from an acceleration of the earth's spin with respect to ephemeris time.

If we let ω_e denote the angular velocity (spin) of the earth with respect to the mean equinox of date,‡ it is natural to use $\dot\omega_e$ for the spin acceleration. However, if we take the units to be seconds of arc and centuries of time, $\dot\omega_e$ has an awkward size. Instead, I introduce a parameter y defined by

$$y = 10^9(\dot\omega_e/\omega_e). \qquad (I.4)$$

The unit of y is the reciprocal century (cy^{-1}). I will usually omit the statement of the units.

We now have four quantities to deal with that relate to the astronomical accelerations. They are $\nu_S{}'$ and $\nu_M{}'$, the accelerations of the sun and moon with respect to solar

---

†The tabulated results also include the rectangular components of velocity, which I do not use in this work.

‡In principle, we should use the angular velocity of the earth with respect to the fixed stars, but it is easier to deal with the equinox. The two rates differ by about 1 part in 10^7.

time, $\dot{n}_M$, the acceleration of the moon with respect to ephemeris time, and the parameter y that was just defined. These quantities are related as follows:

$$\nu_M' = \dot{n}_M - 1.7373\underline{y},$$
$$\nu_S' = \phantom{\dot{n}_M} - 0.1300\underline{y}. \qquad (I.5)$$

The numerical coefficients in Equations I.5 are valid only in the specific set of units that has been adopted.

The numerical integration described in the preceding section does not refer to the earth's rotation, and therefore, it makes no direct contribution to the study of the rotation. In fact, a body of observations that only extends over fifteen or twenty years cannot make much direct contribution to the study of the rotation, regardless of its accuracy, for an important geophysical reason. The work of Spencer Jones [1939], Brouwer [1952], Martin [1969], and Markowitz [1970] has shown that the earth's spin is subject to sudden changes. If we look at the power spectrum of the length of the day [Lambeck, 1980, p. 4, pp. 73ff], we find important components with periods ranging from less than a month to a century or more. The fluctuations with periods from a few years to a century are often called the "decade fluctuations".

There has been much debate about whether the decade fluctuations are best represented by sudden changes in the spin rate, in the spin acceleration, or perhaps by some other model. This question does not need to concern us here. However, in order to gain some idea of the importance of the decade fluctuations, let us follow Markowitz [1970] and postulate that the acceleration does the sudden changing. We find that the interval between changes is fairly random, with an average value of about 4 years. The change in y is also fairly random, and the average size of the change is about 350. In contrast, the average value of y over the past 2000 years is only about -20.

The decade fluctuations in y put a serious restricton on our ability to study the secular component of y. I have shown [Newton, 1973] that the mathematical expectation of the average of the decade fluctuations is about 1 when the averaging interval is about 3 centuries, and I have verified this [volume 1, p. 16] by averaging the actual data over intervals ranging from about 60 years up to 3 centuries. If we accept 1 as a reasonable limit on the accuracy with which we can find y, then we must use data extending over 3 centuries in order to find the secular acceleration. Thus all the data obtained since the invention of the telescope and pendulum clock barely suffice to give us a single value of the secular part of y, in spite of their precision.

In analyzing many ancient and medieval data, we must be able to calculate the Greenwich hour angle of the equinox or, what is the same thing, the right ascension of the Greenwich meridian, as a function of time. Let me call this function the ephemeris of the earth's rotation, or the rotational ephemeris for brevity. We can use either solar time or ephemeris time as the independent variable in the rotational ephemeris, although solar time is the more natural variable.

The rotational ephemeris is inherently based upon solar time but the DE102 ephemeris is inherently based upon ephemeris time, and we must be able to relate the two ephemerides. There are two ways of doing so: (1) We assume a reference meridian called the ephemeris meridian which rotates at a constant rate with respect to ephemeris time, and which is therefore not fixed with respect to the earth. We must then estimate the position of the ephemeris meridian with respect to the meridian of Greenwich as a function of time. (2) We calculate the position of the Greenwich meridian as a function of solar time (using a constant rate of rotation to do so), and we must then estimate the difference between ephemeris time and solar time. I have chosen to use the second approach, which is, of course, equivalent in concept to the first.

Let α_G denote the right ascension of the Greenwich meridian, which is the dependent variable in the rotational ephemeris. In calculating α_G, we start from R_U, the right ascension of the fictitious mean sun. R_U is given as a function of solar (universal) time on page 73 of the Explanatory Supplement [1977], with the reference epoch having Julian day number 2 415 020.0. To start with, let t denote time in Julian centuries from the reference epoch of the rotational ephemeris (whose epoch has yet to be found), and let us change R_U to be a function of t rather than of time from the epoch 2 415 020.0.

When it is noon at Greenwich, α_G equals R_U. At any other hour h, measured from midnight, we add $15°(h - 12)$ to R_U to get α_G in degrees.

The expression for R_U comes from Newcomb's solar theory, which was derived on the tacit assumption that the sun has no acceleration with respect to solar time.† If it has an acceleration ν_S', the value of R_U must be increased by $\frac{1}{2} \nu_S' t^2$ seconds of arc. If we continue to let R_U denote the specific function that is given in the Explanatory Supplement, the complete expression for α_G as a function of solar time is:

$$\alpha_G = R_U + 15°(h - 12) + \frac{1}{2} \nu_S' t^2,$$

----

†There is a term in Newcomb's theory that is quadratic in the time, but it arises solely from planetary and lunar perturbations. It does not arise from an acceleration in the sense that we use the word here.

with the last term being in seconds of arc. However, we
have agreed to base the work of this volume primarily on
ephemeris time rather than solar time, so we should use the
parameter $\underline{y}$ rather than ν_S'. Thus, if we substitute
for ν_S' from the second of Equations I.5, we finally have

$$\alpha_G = \underline{R}_U + 15°(\underline{h} - 12) - 0''.0650\underline{y}t^2. \qquad (I.6)$$

Besides the rotational position of the earth, we will
have to deal with the positions of the sun, the moon, the
planets, and the stars. The numerical integration of the
positions in the solar system did not have anything to do
with the positions of the stars. Thus, I will calculate the
positions of the stars in the way I have always done, except
for using the new value of the constant of precession.

In order to enter the DE102 ephemeris, we must have a
value of ephemeris time. However, the values of time that
come from the ancient and medieval data are in solar time,
and so is the independent variable in the rotational ephemeris. Hence, in order to use the DE102 ephemeris, we must
have a value of ΔT, the difference between ephemeris time
and solar time. The question of choosing $\underline{\Delta T}$ is closely
connected with the question of finding the reference epoch
of the rotational ephemeris, and I will take it up in a
moment. In the meantime, let us assume that $\underline{\Delta T}$ is known and
see how we use it.

We add $\underline{\Delta T}$ to a value of solar time to obtain a value of
ephemeris time, and we use this value of ephemeris time to
enter the DE102 ephemeris. If we are dealing with the sun
or the planets, we make no change in the positions given by
the ephemeris. For the moon, however, we assume that it has
an acceleration $\underline{\dot{n}}_M$ with respect to ephemeris time, and thus
we do not necessarily use the position given by the ephem-
eris.

I noted in the preceding section that $\underline{\dot{n}}_M$ is taken to be
-26.21 in the numerical integration that leads to the DE102
ephemeris. However, we must allow for the possibility that
this value is not necessarily correct. That is, we must
admit a new parameter, that I will call $\underline{x}$, defined as

$$\underline{x} = \underline{\dot{n}}_M + 26.21 \qquad\qquad ''/cy^2. \qquad (I.7)$$

Thus, after we find the longitude of the moon from the DE102
ephemeris, we add $\tfrac{1}{2} \underline{x}T^2$ to it.†

We now have two acceleration parameters, $\underline{\dot{n}}_M$ and $\underline{y}$, both
referred to ephemeris time. In such a situation, we usually
vary both parameters until we find the best fit to the data.

---

†$\underline{T}$, we must remember, is measured from the epoch of the
DE102 ephemeris, which has the Julian day number
2 440 400.5.

Here, however, for a reason that will be explained in Section I.7 below, I assume a value for $\dot{n}_M$ and find only the parameter y from an analysis of the data. In the present section, it remains to explain how we find the reference epoch of the rotational ephemeris and how we find ΔT, which means ephemeris time minus solar time.

Ephemeris time is "officially" defined as the independent variable in the equations of motion of the sun, moon, and planets, but this is an expression of intent rather than of observational fact. The actual measure of ephemeris time to which atomic time was matched was determined from the motion of the moon, which is significantly affected by tidal friction.

Specifically, let us suppose that all perturbations in the motion of the moon due to the gravitation of the sun and planets (and to the nonsphericity of the earth) are adequately removed by the theory of Brown [1919], as modified by Eckert, Jones, and Clark [1954], so that an observation of the lunar position gives us an "observed" value L_{obs} of the moon's mean longitude. If there were no other perturbations, L_{obs} would be a linear function of time, within observational error. If we let S denote solar time measured from noon on 1900 January 0,† we find that L_{obs} is not a linear function of S. Instead, we find that L_{obs} differs systematically from a linear function, and to handle the situation we write:

$$L_{obs} = L_O + \dot{L}S + \delta L. \tag{I.8}$$

The quantity δL comes from two main sources, in addition to the relatively small errors of observation. One source is tidal friction, whose effect is close to a quadratic function of time over a few centuries. The other source is the random fluctuation of the earth's spin that was discussed earlier in this section.

Brouwer [1952] and Martin [1969] have made extensive determinations of δL, and I tabulated the 5-year averages of δL obtained from their work from 1627 to 1957 on page 21 of volume 1.‡ In order to eliminate the effect of the large random (decade) fluctuations, we must use this entire interval. The quadratic that best fits the values of δL over the interval stated is (volume 1, p. 19)

$$-3''.79 - 14''.48S - 6''.31S^2.$$

If we ignore the effect of the random fluctuations, then, we can write

---

†It is necessary to use 1900 as a reference epoch in this historical discussion.

‡In the first part of this time interval, the data are not sufficiently dense to allow finding 5-year averages, and longer averages were used.

$$\underline{L}_{obs} = \underline{L}_o + \dot{\underline{L}}S - 3''.79 - 14''.48\underline{S} - 6''.31\underline{S}^2. \qquad (I.9)$$

Now let E denote ephemeris time from 1900 January 0 at noon.† The work of Brouwer, Martin, and others has led to taking

$$-8''.72 - 26''.74\underline{E} - 11''.22\underline{E}^2$$

as the effect of tidal friction on the motion of the moon in ephemeris time. That is, we take

$$\underline{L}_{obs} = \underline{L}_o + \dot{\underline{L}}E - 8''.72 - 26''.74\underline{E} - 11''.22\underline{E}^2. \qquad (I.10)$$

The constants $\underline{L}_o$ and $\dot{\underline{L}}$ are supposed to represent the purely gravitational orbit of the moon, and they are the same whether we use solar time or ephemeris time.

We let ΔT denote the difference between E and S:

$$\Delta T = \underline{E} - \underline{S} = \text{ephemeris time - solar time.} \qquad (I.11)$$

We can find $\Delta\underline{T}$ by combining Equations I.9 and I.10. The result is:‡

$$\Delta\underline{T} = 8^S.98 + 22^S.33\underline{E} + 8^S.94\underline{E}^2. \qquad (I.12)$$

ΔT in Equation I.12 has a minimum value of $-4^S.96$ when E equals -1.25. This is the year 1775. Even though the correction for tidal friction in Equation I.10 is probably not the correct one, nonetheless it is the one that was used in establishing the scales of ephemeris time and atomic time (and thus dynamical time also), and hence Equation I.12 is the one we should use for ΔT. However, it is interesting to see how sensitive $\Delta\underline{T}$ is to observational and analytic uncertainties.

In Section I.4 of volume 1, I made a new analysis of the data which led to Equation I.10, eliminating some of the approximations that had been made in the earlier analysis. I found that $\dot{\underline{n}}_M$ should be -28.38 rather than -22.44, so that the coefficient of E^2 in Equation I.10 should be -14.19 rather than -11.22. I failed to preserve the other coefficients in the quadratic that represents tidal friction, but it is easy to find a satisfactory approximation for them. We ask what function of the form $\underline{a} + \underline{b}E - 14''.19\underline{E}^2$ is the

---

† I remind the reader that noon on 1900 January 0, ephemeris time, is not the same epoch as noon on 1900 January 0, universal (solar) time.

‡ In order to let ΔT be expressed in seconds, we take $\dot{\underline{L}}$ in Equations I.9 and I.10 to be in "/sec instead of the conventional "/cy. The value of $\dot{\underline{L}}$ is 0.549 02 "/sec. When we write Equation I.12, we equate E and S after we solve for $\Delta\underline{T}$. Since $\Delta\underline{T}$ is a small quantity, it does not matter whether we use ephemeris time or solar time in evaluating it from Equation I.12.

best approximation to $-8''.72 - 26''.74E - 11''.22E^2$. The answer is

$$-8''.84 - 31''.19E - 14''.19E^2. \tag{I.13}$$

When we use this quadratic in Equation I.10 instead of the one that appears there, we get

$$\Delta T = 9^S.20 + 30^S.44E + 14^S.35E^2. \tag{I.14}$$

This has a minimum value of $-6^S.94$ when E equals -1.06. This is the year 1794.

When we analyze the old data, we will find new values of the coefficient of E^2, so the difference in the coefficient of E^2 between Equations I.12 and I.14 does not matter. If we refer ΔT to the minimum, the difference in the coefficients of E does not matter, because the linear coefficient goes to zero. Only the difference in the minimum values, and in the epoch when the minimum occurs, affect the situation. The difference between 1775 and 1794 is unimportant when we are examining data obtained before 1600, and the difference between $-4^S.96$ and $-6^S.94$ is also unimportant.

For convenience, I take the reference epoch for ΔT to be Julian day number 2 375 000.5. This is the midnight that began 1790 June 6, Gregorian calendar. By implication, this is also the reference epoch for the rotational ephemeris of the earth. I take the value of ΔT at the reference epoch to be -5 seconds, and I assume that the reference epoch is the minimum of ΔT.

Let t denote time in Julian centuries from the epoch 2 375 000.5. Then I take ΔT to be

$$\Delta T = -5 + ct^2,$$

in which c is a constant that is directly proportional to the acceleration parameter y. From Equation I.28, page 32 of volume 1, we find that $y = -0.6338c$. Hence

$$\Delta T = -5 - 1.5778yt^2 \quad \text{seconds.} \tag{I.15}$$

We can now summarize the general procedure for using the various ephemeris programs. An observation that is to be used has an associated solar time t. With this value of solar time, referred to Julian day number 2 375 000.5, I calculate the right ascension of Greenwich at the time of the observation from Equation I.6, using a conventional value of y. I also calculate a value of ΔT from Equation I.15 using the same value of y.

I then add this value of ΔT to the solar time of the observation, and refer this new time to Julian day number 2 440 400.5. This gives the value of T used to enter the DE102 ephemeris. The positions from this ephemeris, after they are transformed to the coordinate system of date instead of to the fixed system of 1950.0, are left unaltered, except for the longitude of the moon. It is increased by

$\frac{1}{2}\underline{x}t^2$, where $\underline{x}$ is related to the acceleration of the moon $\underline{\dot{n}}_M$ by Equation I.7.

5. A Summary of Volume 1

In volume 1, I analyzed the modern data obtained since the introduction of the telescope and pendulum clock to find the mean motions and the secular accelerations of the sun and moon. I also studied about 2000 historical texts before the year 1600 that record observations of solar eclipses. I confined this study to what we may call "amateur" observations. By this I mean simple statements, usually by people with no training in astronomy, that a solar eclipse had been observed at a known place on a known date. Sometimes these statements are accompanied by approximate statements about the hour of the day or the magnitude of the eclipse. I reserved for volume 2 the relatively few attempts, usually by people with some training, to measure the hour or magnitude accurately, and I did not use these attempts in volume 1. I used only the fact that an eclipse had been observed at a known place and time.†

From the total number of eclipse records studied, I selected 631 which, on the basis of textual analysis, were independent, could be dated, and could be assigned to a particular point of observation,‡ all with reasonable assurance. The dates of the 631 records used range from -719 February 22 to 1567 April 9.

I analyzed these records using solar time as the time base. That is, I took the time base to be defined by the rotation of the earth, so that the earth's spin has no acceleration. I assumed that the sun and moon have accelerations with respect to solar time that are called $\nu_S{}'$ and $\nu_M{}'$, respectively; these are related to $\underline{\dot{n}}_M$ and $\underline{y}$ by Equations I.5. In principle, both $\nu_S{}'$ and $\nu_M{}'$ are determined by the observations, but it works out that they are determined weakly. Only one parameter is determined strongly by the eclipse observations, and it is the parameter that I called D''. This is the second derivative of the moon's mean elongation $\underline{D}$ from the sun, and it equals $\nu_M{}' - \nu_S{}'$.

---

†However, if a record says that an eclipse was very large or total, I gave it a higher weight than one which merely records the occurrence of an eclipse. This is not because such a record is more reliable than a simple record of an eclipse. Rather, since an eclipse is large or total only in a narrow zone, an assertion of a large or total eclipse imposes tighter constraints on the parameters.

‡In some cases, particularly with the earlier records, it is sufficient to know only that the point of observation was somewhere within a fairly small region, say a region the size of Bavaria.

In contrast, the modern data studied in volume 1 give a strong determination of both $\nu_M{}'$ and $\nu_S{}'$, and the value of D'' calculated from these values joins smoothly with the values of D'' found from the older eclipse observations. Thus, volume 1 gives us a consistent picture of D'' from, say, -700 to the present. This picture is as follows:

D'' is a strong function of time which was positive before about the year 500 and which has been negative ever since. When we represent D'' as a power function of the time, we should use the center of the data as the origin of time. This center is at about +600. I use C to denote time in centuries measured from 600. If we find the best linear fit to the values of D'' we get (volume 1, p. 447):

$$D'' = -1.54 \pm 1.44 - (0.854 \pm 0.155)C. \qquad (I.16)$$

The best quadratic fit is:

$$D'' = +0.58 \pm 2.55 - (0.844 \pm 0.156)C$$
$$- (0.0244 \pm 0.0243)C^2. \qquad (I.17)$$

The coefficient of C^2 in Equation I.17 barely has significance from the statistical point of view, but there are good physical grounds, which I discuss in Section XIV.9 of volume 1, for believing that the quadratic term should be present. The coefficient of C in the two equations is about the same, and it is more than five times its estimated standard deviation, so it is well established that D'' is a strong function of time.

I also looked at the accelerations $\dot{n}_M$ and y in volume 1, even though they are weakly determined by the eclipse data. However, they are strongly determined by the modern data, with the result that

$$\dot{n}_M = -28.38 \pm 5.72 \qquad (I.18)$$

at the present time. From the eclipse data I got (Equation XIV.9, p. 455, volume 1)

$$\dot{n}_M = -29.41 \pm 21.00, \qquad (I.18a)$$

when evaluated at the epoch $C = 0$. The eclipse data also yield a small time dependence that is too small to be statistically significant.

The two estimates in Equations I.18 and I.18a agree better than we have any right to expect, even though they are evaluated at epochs 12 centuries apart. Of course, the standard deviation in Equation I.18a is large, and we cannot exclude some time dependence of $\dot{n}_M$. If there is a time dependence, however, it is much less relatively than the time dependence of D''. Further, though there are good physical grounds for believing that D'' is a strong function of time, there are good grounds for believing that $\dot{n}_M$ has been substantially constant, at least during a period as short as that from −700 to the present.

I discussed the physical bases for the changes in D'' and the constancy of $\dot{n}_M$ in Section XIV.9 of volume 1.

If we adopt the values of D'' and $\dot{n}_M$ from Equations I.17 and I.18, and solve for y from Equation I.22 below, we find

$$y = -9.62 + 0.8862t + 0.015\ 154t^2. \qquad (I.19)$$

In this, to repeat, t is time in Julian centuries from the epoch with the Julian day number 2 375 000.5. This is in the year 1790. I adopt this value, and the value of $\dot{n}_M$, as standard values to be used in discussion, pending the derivation of new values in this volume.†

6.   The Purpose of Volume 2; Types of Data to Be Used

In volume 1, I used two types of data. One type consists of observations made since the introduction of the telescope and pendulum clock into observational astronomy. The other type consists of what I called "amateur" observations of solar eclipses a moment ago.

The purpose of volume 2 is to see what we can learn about the accelerations by using "professional" observations, whether of solar eclipses or of other phenomena, that were made before the introduction of the telescope and pendulum clock. By "professional" observations, I mean those made by people who were well trained in astronomy, whether or not they depended upon astronomy for an income. I will not introduce any additional observations made with the telescope and clock. It is likely that the use of additional data will contribute usefully to the study of the decade fluctuations, but it is not likely that the additional data now available will add much to our knowledge of the secular accelerations.

Almost all the data to be used in this volume come under one of the following classes:

   a.   Time intervals between moonrise or moonset and sunrise or sunset. Most of these data are Babylonian and were measured between −567 and −232.

   b.   Measured times of lunar eclipses.

   c.   Measured times of solar eclipses.

   d.   Measured magnitudes of lunar eclipses.

   e.   Measured magnitudes of solar eclipses.

---

†Actually, I will round the value of $\dot{n}_M$ to −28. See Section I.7 below.

f. Timed conjunctions of the moon with stars or
    with the planets.

g. Timed occultations of stars or planets by the
    moon.

h. Simultaneous measurements of the azimuth and
    elevation of the moon.

i. Measured time intervals between the meridian
    transit of the moon and of a star or planet.

Observations of classes d, e, and h do not involve the
measurement of time, but all the other classes do. In some
cases, I have described the classes as directly involving
the measurement of a time interval. If this interval is
reasonably short, say half an hour, the measurements should
be reasonably accurate even for observations as old as the
Babylonian ones.

If an observation involves the measurement of epoch
time, as opposed to an interval of time, I use it only if
the time was measured by some astronomical means and not by
a clock. Various acceptable ways of measuring the time
include measuring the azimuth or elevation of the moon itself at the time of the measurement, or of measuring the
azmuth or elevation of some other body.

7. General Remarks About the Analysis

Since many different types of data are involved in the
present volume, it is not feasible to describe their analysis in one place. However, there are some general remarks
about the principles of analysis that can usefully be made
here.

There needs to be one main change from the principles
of analysis that were used in volume 1. There, I adopted
standard expressions for the mean longitude $\underline{L}_{\underline{M}}$ of the moon
and the mean longitude $\underline{L}_{\underline{S}}$ of the sun. These, by subtraction, gave a standard expression for the mean elongation $\underline{D}$
of the moon, where

$$\underline{D} = \underline{L}_{\underline{M}} - \underline{L}_{\underline{S}} \, . \qquad\qquad (I.20)$$

I then calculated the circumstances of a particular eclipse
using this standard value of D and other values close to it
until I found the value of D that gave best agreement with
the observed circumstances. Let us write this value of D in
the form D + δD, where D still means the value from Equation
I.20. Then, if the eclipse occurred τ centuries before the
reference epoch, I defined an average acceleration parameter
$\underline{D}''$ by setting

$$\delta\underline{D} = \tfrac{1}{2} \, \underline{D}'' \tau^2 \, . \qquad\qquad (I.21)$$

Proceeding this way with eclipses from different periods of

history, I found D'' as a function of time, in the manner described in Section I.5 above.

Using the various accelerations that have been introduced earlier, we related $\underline{D}''$ to them by:

$$\underline{D}'' = \nu_M{}' - \nu_S{}' = \underline{\dot{n}}_M - 1.6073\underline{y}. \qquad (I.22)$$

We could do this because all the accelerations in the right members of Equations I.22 had the same reference epoch, so that τ had the same value for all of them.

Now we may no longer do this; there is no value of τ that applies to both accelerations. The acceleration $\underline{\dot{n}}_M$ of the moon with respect to ephemeris time is referred to the reference epoch 1969 while the acceleration of the earth's spin is referred to the reference epoch 1790, almost 180 years earlier. Thus, for physical reasons, there is no value of τ to use in Equation I.21 and thus a value of δD no longer leads to a value of D''. In fact, we can no longer define values of $\nu_M{}'$ and $\nu_S{}'$; we can only define the individual accelerations $\underline{\dot{n}}_M$ and $\underline{y}$.

Of course, we can still formally define $\nu_M{}'$ and $\nu_S{}'$ by Equations I.5 if we have values for $\underline{\dot{n}}_M$ and $\underline{y}$. However, we can no longer use $\nu_M{}'$ and $\nu_S{}'$ to calculate perturbations in $\underline{L}_M$ and $\underline{L}_S$, just as we can no longer use Equation I.21. This is the basic reason why I have gone to the use of ephemeris time in this volume.

In principle, we can find both $\underline{\dot{n}}_M$ and $\underline{y}$ by studying lunar observations. However, whenever I have tried to do so, I have found that the simultaneous equations for the two parameters are nearly singular, so that there are large uncertainties in the solutions. Thus, only one parameter can be found with useful accuracy. In this volume, I will not even try to find two parameters. The one that I shall find is $\underline{y}$.

Specifically, I assume a standard value for $\underline{\dot{n}}_M$. With this value of $\underline{\dot{n}}_M$, I calculate an observed quantity with some convenient value of $\underline{y}$ and also with a neighboring value of $\underline{y}$. This will allow me to find a derivative with respect to $\underline{y}$. With this information, I can then find the value of $\underline{y}$ that best fits an observation or a set of observations.

In volume 1 (page 457), I reached the estimate that $\underline{\dot{n}}_M$ = -28.38 ± 5.72. I do not know when the custom arose of giving the accelerations to a precision of four decimal positions when the value is uncertain by several units in the second position, but this practice has been followed in much of the literature for a long time. I will break with custom and retain only two decimal positions in $\underline{\dot{n}}_M$. Thus, I take the standard value of $\dot{n}_M$ to be

$$\dot{n}_M = -28. \qquad (I.23)$$

It will turn out that we know y to an accuracy of about 1 unit in the second place, if the standard value of $\dot{n}_M$ is correct. That is, $\dot{n}_M$ and y are highly correlated, and we know y for a given value of $\dot{n}_M$ much more accurately than we know either value individually. In order not to lose this correlation, I will give values of y to three decimal positions.

Although I will not use them in this volume, I will calculate and tabulate the derivatives of each observed quantity with respect to $\dot{n}_M$. This will help the reader who wants to make his own analysis of the observations using some other value of $\dot{n}_M$.

The analysis of the "amateur" observations in volume 1 was based upon the older methods, so the results found there cannot be combined easily with the ones to be found here. As a consequence, I have made a new analysis of the data from volume 1 along the lines just indicated. The results are found in Appendix I, and it is these results that will be combined with the results from the "professional" data to be presented in this volume. The difference between Appendix I and volume 1 is not particularly important; it is just large enough that we need to take it into account.

8.   The Mean Longitudes

I mentioned in Section I.3 that Van Flandern [in preparation] has derived new mean elements for the sun and moon from the DE102 ephemeris, and I gave his expression for the obliquity ε in Equation I.3. The only other ones of his elements that we need explicitly in this volume are the mean longitudes L_S and L_M of the sun and moon, respectively. Van Flandern gives them referred to the epochs with Julian day numbers 2 415 020.0 (1900 January 0.5) and 2 451 545.0 (2000 January 1.5),† but we need them referred to the Julian day 2 440 400.5, which is the prime epoch for the DE102 ephemeris, and the epoch to which the acceleration of the moon is referred. If we let T denote time from day 2 440 400.5, expressed in Julian centuries of 36525 days, we have

$$L_S = 95°.919\ 102 + 36\ 000°.769\ 613T + 0°.000\ 302\ 5T^2. \quad (I.24)$$

The expression for L_S is independent of the value used for $\dot{n}_M$, but the expression for L_M is not. Van Flandern uses

---

†Van Flandern chose the epochs to be separated by one Julian century. This means that they are separated by one day more than the Gregorian century, since 1900 was not a leap year.

$\dot{n}_M$ = - 26.21, but I have adopted -28 in this volume. Van Flandern's expression for L_M is:

$$L_M = 253°.965\ 758 + 481\ 267°.882\ 262T$$
$$- 0°.001\ 616\ 5T^2 + 0°.000\ 005\ 28T^3.$$

If we take $\dot{n}_M$ = -28, we must subtract $(1.79/7200)T^2$ (degrees) from Van Flandern's value. Doing so gives

$$L_M = 253°.965\ 758 + 481\ 267°.882\ 262T$$
$$- 0°.001\ 865\ 1T^2 + 0°.000\ 005\ 28T^3. \qquad (I.25)$$

We will also have frequent need for the mean elongation D of the moon from the sun, which is defined as $L_M - L_S$. Hence,

$$D = 158°.046\ 656 + 445\ 267°.112\ 649T$$
$$- 0°.002\ 167\ 6T^2 + 0°.000\ 005\ 28T^3. \qquad (I.26)$$

Equation I.26, like Equation I.25, is based upon the value $\dot{n}_M$ = -28.

9.   Notation

There is not much notation in this volume, and most of that little occurs only locally. It is probably not worth-while to list such notation, because it is always defined close to the place where it is introduced. However, there are a few symbols which are used throughout the volume, and they are listed here.

C :   time in Julian centuries from the year 600 (C is used only in expressions of low accuracy for which a precise epoch is not needed.)

D :   mean elongation of the moon

L_M :   mean longitude of the moon

L_S :   mean longitude of the sun

m :   magnitude of the diameter of an eclipse

$\dot{n}_M$:   acceleration of the moon with respect to ephemeris time

$\underline{t}$: time in Julian centuries from the epoch
       2 375 000.5

$\underline{T}$: time in Julian centuries from the epoch
       2 440 400.5, except in the combination $\Delta\underline{T}$

$\underline{y}$: the earth's spin acceleration, in parts in
       10^9 per century, with respect to ephemeris
       time

$\Delta\underline{T}$: ephemeris time minus solar time, in seconds

$\omega_{\underline{e}}$: the earth's spin rate, in seconds of arc per
       century

$\dot{\omega}_{\underline{e}}$: the earth's spin acceleration, in seconds of
       arc per century per century, with respect to
       ephemeris time

    When I write a number in sexagesimal notation, I sepa-
rate the integral and fractional parts by a semicolon.
Within the integral and fractional parts, I separate the
various positions by commas. Thus,

$$5,29;59,23,17 = 5 \times 60^1 + 29 \times 60^0$$

$$+ 59 \times 60^{-1} + 23 \times 60^{-2} + 17 \times 60^{-3}.$$

CHAPTER II

LUNAR OBSERVATIONS MADE BY TWO RENAISSANCE ASTRONOMERS

1. Tycho Brahe

In this chapter I will take up the observations of the
moon made by two Renaissance astronomers, namely Tycho Brahe
and Copernicus, to put them in inverse chronological order.
For various reasons, the observations made by these astronomers can contribute so little to the study of the moon's
acceleration that they are not worth the effort of analysis.
Nonetheless, because of the stature of these men, I think it
is worth spending some time in explaining why their observations are not valuable for the purposes of this work.   I
start with Tycho Brahe.

Tycho has left us a long series of systematic astronomical observations, numbering into the tens of thousands,
which were probably the most precise observations ever made
before the invention of the telescope. For this reason, I
originally thought that his observations would be quite
useful, in spite of their relatively recent date. This
would be so if it were a matter only of the precision of the
observations. Unfortunately, as we will see, Tycho's observations contain systematic errors that cannot be reduced to
an acceptable level by using large quantities of data. In
order to be useful, his observations would need to be at
least two centuries earlier than they are.

I will start the discussion of Tycho's observations
with a short description of his life and career. The book
by Dreyer [1890], in spite of its age, is still probably the
best single source on Tycho. The article by Hellman [1970]
is a good short summary, and it gives additional references.

Tycho was born to a noble and fairly wealthy family who
lived in what is now southern Sweden but was then part of
the kingdom of Denmark. He was born in 1546, the same year
that Martin Luther died. The religious dissension of the
time had an important effect upon Tycho's life and career,
and it probably had a greater effect upon Tycho's assistant
Johannes Kepler.

Tycho's family required him to study law, but he acquired an interest in astronomy at an early age. As a law
student, he spent all his spare money in buying astronomical
instruments and all his spare time in using them to make
astronomical observations. By the time he was 30, he had
already acquired considerable reputation as an astronomer.
About this time, he also became a successful practitioner of
grantsmanship.

In 1576, King Frederik II of Denmark gave Tycho a royal
grant, which was continued annually for many years, and
which Tycho used to build a magnificent home and two observatories on the island of Hveen in the Kattegat about midway
between the present Denmark and Sweden. Tycho equipped his

observatories with instruments of his own design which were probably the most advanced astronomical instruments that had ever been built. He was also able to support a considerable number of assistants. Some of his assistants are named explicitly in the logbook of his observations. In his observatories on Hveen, Tycho and his assistants made the largest and most systematic series of astronomical observations that the world had yet known.

Frederik II died in 1588 and was succeeded by Christian IV. Tycho's relations with Christian were not as cordial as they had been with Frederik, and in 1596 Christian began to reduce the size of Tycho's annual grants. As a result, Tycho left Denmark forever in the spring of 1597, and eventually settled at the court of the Holy Roman emperor.† He died there in 1601.

Tycho continued to make observations after he left Hveen, but he never again had a settled place in which he could make a long series of observations on a homogeneous basis with elaborate instruments. In this volume, therefore, I consider only the observations that he made on Hveen. These observations were made between December of 1576 and the spring of 1597.

In addition to writing his biography, Dreyer [1926] has prepared an edition of the collected works of Tycho, including the logbook of the observations made in his observatories. These include not only the observations made by Tycho himself but also the observations made in his observatories by his many students and assistants. The observations take up volume X through volume XIII of the collected works. They are arranged by year. Under each year, they are systematically arranged according to the body being observed. Because of this systematic arrangement, I do not believe that it is necessary to cite the volume and page where any particular observation appears.

I will not distinguish between the observations made by Tycho himself and those made in his observatories by his assistants under his general supervision.

---

†There are several points of interest about Tycho's relations with Christian. The large early grants to Tycho were for the purpose of supporting the construction of his observatories, and such large grants were not needed for their maintenance and operation. It is likely that Tycho left Hveen as an enormously wealthy man, far richer than when he moved there. As I understand the matter, Christian could not cut off Tycho's grants entirely, because Frederik had granted him the total income from the island of Hveen for Tycho's life, and he continued to receive these revenues even after he entered the service of the emperor. All things considered, Christian was probably justified in decreasing but not eliminating Tycho's grants of income. Tycho seems to have been a mixed and complicated man.

2.   The Accuracy Needed from Tycho's Observations

The observations made on Hveen have an average date of about 1590, and the reference epoch for the earth's rotation (Section I.4) is 1790.   Thus the "lever arm" of Tycho's data is 2 centuries, so they must give us the position of the moon with considerable accuracy if they are to make a significant contribution to the determination of $\underline{y}$.†

In Section XIV.1 of volume 1, we concluded that we had found $\underline{D}''$ to a precision of about 2 $''/cy^2$ (standard deviation) at any time since about -700.   From Equation I.22 above, we see that an uncertainty of 2 in D'' is equivalent to an uncertainty of 1 ¼ in $\underline{y}$.   From Equation I.15 above, we see that this uncertainty in $\underline{y}$ is equivalent to an uncertainty of about 8 seconds in the value of $\Delta\underline{T}$ for an observation made 2 centuries before 1790.   In other words, if Tycho's observations are to be useful for our purposes, they must give us $\Delta\underline{T}$ to an accuracy of about 8 seconds.

During this time interval, the moon moves about 4".   Thus we need to get a position to about 4 seconds of arc and a time to about 8 seconds of time if Tycho's observations are to be useful.   He was able to measure both position and time to a precision of about 30 seconds.   Thus we can reach the needed precision with about 16 × 64, or about 1000, measurements.   Tycho made several thousand observations of the moon, and so we can get a useful estimate of $\underline{y}$ from his data if there are no systematic errors in them large enough to interfere.

If there are systematic biases in his observations, of a nature that cannot be removed by analysis, they must be no more than 4" in angular position or 8 seconds in time.   These are rather strict limits for observations made with the naked eye and without a pendulum clock.‡

3.   Types of Lunar Observation

To start with, Tycho systematically observed all the eclipses of the sun and moon that were visible on Hveen,‡ weather permitting.   He estimated the times of the various

---

†Further, 1590 is about as late as we can come in time without worrying about the "decade" fluctuations in the earth's spin acceleration.   See Section I.4.

‡That is, without one regulated by a gravity pendulum. Galileo discovered the principle of the gravity pendulum during the time that Tycho was observing on Hveen, but the gravity pendulum, according to the standard histories, was first applied to a clock by Christian Huyghens in 1656. However, the torsion pendulum, with an escapement, had been available for more than two centuries.

‡Some of the eclipses are logged under the solar observations and some under the lunar ones.

phases of the eclipses, and he estimated the magnitude as a
function of time. It seems to me that the times are necessarily biased. He could know that a solar eclipse, say, had
begun only when the moon had moved a finite distance across
the sun, and he could know that it had ended only when the
moon definitely no longer obscured any part of the sun.
Thus, as it seems to me, the times are biased toward being
late.† Put another way, the position of the moon relative
to the sun at the time of beginning or ending of an eclipse
is biased.

I believe that the people with the best vision can
resolve no more than 5″ under such circumstances, and the
resolution is poorer for most people. Even if 5″ is correct, however, this exceeds the bias that we can allow, and
I do not see any way to estimate the bias and to remove it
from the data. Thus it is not safe to use measurements of
times associated with eclipses.

I do not know enough about the process of estimating
the magnitude of an eclipse with the naked eye to feel safe
about using Tycho's estimates of the magnitude. Thus I will
not use any of Tycho's observations of eclipses.

Now let us turn to Tycho's measurements of time in
general. He had several clocks, and he sometimes gives the
results of simultaneous readings of as many as three of
them. Sometimes he makes a systematic set of observations
using two of the clocks, naming them and reading both for
every observation. Sometimes he uses only one clock and
names it. Most often, though, he uses only one clock without naming it.

He does not tell us much about his clocks, and we do
not know how they were regulated. All we can say is that
they were not regulated with a gravity pendulum, as I have
already remarked in a footnote. In using a clock, Tycho
usually set it to read noon at the transit of the sun, so
that it read apparent local time. Then, at the next noon,
weather permitting, he read and recorded the time of the
solar transit and then reset the clock. In addition, he
measured the times of many stellar transits, and occasionally he reset his clock by one of them. Usually, though,
he let a clock run freely until his next opportunity to set
it by the sun.

In a forthcoming study, I analyze a large number of
measured times of transits of the sun and of stars and study
the errors in them. I find that the errors are dominated by
a random walk that has a standard deviation of about 15
seconds for an averaging time of 10 minutes. Now suppose
that Tycho measured the position of the moon at a time 8
hours after a noon when he had set his clock by the sun. At

---

†Of course, Tycho could have corrected for this effect, as
 ibn Yunis [1008] tried to do. (See Section V.3 below.)
 However, I see no suggestion in his work that Tycho tried
 to make this correction.

this time, we expect a clock error of about $15\sqrt{48} = 105$ seconds, say. If the errors are indeed random, we can reduce the resultant error to 8 seconds by using about 175 observations. Tycho made far more than this number.

Unfortunately, but naturally, Tycho made most of his observations of the moon's position at night, when the temperature was systematically lower than at noon when the clock was usually set. It is almost certain that clocks as primitive as his had a systematic temperature dependence as well as a random walk, and so it is almost certain that there is a bias in his measurements of time that cannot be removed statistically.

I have found only two types of lunar observation made by Tycho that hold any promise of being particularly useful, and both involve finding the time by some means other than a clock. In one method, Tycho measured some coordinate, say the azimuth, of the moon at a stated time on one of his clocks, and he did the same thing for a star or planet within a short time before or after. With these observations, we use the star or planet position to correct the clock, and we then hope that the change in the clock error is negligible during the short time between the two observations.

In the other method, Tycho measured the azimuth and altitude of the moon simultaneously. In this method, we use the azimuth to find the time of the observation and then use the altitude to find the lunar acceleration.

The next thing we must do is study the precision with which Tycho could measure a coordinate of the moon. There is a problem with the moon that does not arise with the sun, with a star, or with a planet, which we will turn to in the next section.

4.   Tycho's Measurements of the Lunar Diameter

We will see in Section II.6 below that Tycho achieved a precision considerably better than 1' in measuring the position of the sun, the stars, and the planets. If the moon were a point source like a star or planet, or if it had circular symmetry like the sun, we would expect the same precision for Tycho's lunar observations. However, the moon is an extended and usually unsymmetrical object, and we have considerable trouble in trying to find what point is its center.

Tycho recognized this problem, and he tried to solve it by measuring the position of the bright limb of the moon rather than by trying to estimate the position of its center. In doing so, however, he created another problem that he probably did not know about: When a person looks at the bright limb of the moon against the dark night sky, optical effects in the eye make the limb appear larger than it really is. Thus, for example, if the east limb is the illuminated one, it appears to be farther east than its true position. Before we try to use Tycho's measurements of the lunar position (or, rather, of one of its limbs), we must study this problem.

TABLE II.1

TYCHO'S MEASUREMENTS OF THE LUNAR DIAMETER

Date	Recorded hour	Diameter Measured '	Calculated '	Residual[b] '
1578 Sep 16	2:52	32	30.03	+0.45
1578 Oct 7	18:10:30	34	31.60	+0.88
1582 Oct 24	19:15	36	31.51	+2.97
1582 Nov 24	17:06	36.07	32.71	+1.84
1583 Jan 21	18:22:45	38.5	32.71	+4.27
1583 Jan 23	20:57:50	34.5	32.52	+0.46
1584 Jul 10	22:29:45	35.5	29.54	+4.44
1584 Aug 8	~22.0[a]	31.5	29.72	+0.26
1584 Sep 8	~23.0[a]	31	30.71	-1.23
1584 Oct 4	~19.8[a]	30	30.22	-1.74
1585 Jan 2	~21.3[a]	32.5	33.12	-2.14
1585 Jan 7	~ 1.3[a]	32.5	33.87	-2.89
1585 Jan 9	~23.8[a]	30.25	32.45	-3.72
1585 Jan 10	~ 3.8[a]	30.25	32.45	-3.72
1585 Jan 14	6:40:30	29.33	30.41	-2.60
1585 Feb 28	19:25:20	38	32.91	+3.57
1585 Mar 30	20:10	28.42	32.91	-6.01
1585 Sep 8	9:27	27	32.66	-
1585 Sep 10	10:03	26	33.38	-
1585 Sep 10	11:06:30	28	33.36	-
1585 Sep 20	17:45	31.5	30.17	-0.19
1585 Sep 20	19:38:40	30	30.13	-1.65
1585 Sep 21	18:55	35	29.90	+3.58
1585 Sep 22	18:13	30.25	29.73	-1.00
1585 Sep 22	19:42	32.5	29.73	+1.25
1585 Sep 23	18:30	30	29.64	-1.16
1585 Sep 23	20:29	28	29.67	-3.19
1586 Jan 22	19:11:20	39	32.90	+4.58
1586 Jan 22	22:31:40	37.25	33.05	+2.68

[a]Tycho did not record the time of this observation. I have estimated the tabulated time from the complete record of the observing session.

[b]The measured value minus the calculated value minus 1'.52.

We can do this by using the measurements that Tycho made of the size of the moon. Tycho measured the apparent diameter of the moon on a number of occasions by measuring the altitude of both its upper and lower limbs. Table II.1 gives the results of 29 such measurements. The first column

in the table gives the date† of the observation and the
second column gives the hour that Tycho recorded, in local
time at Hveen. These columns are not highly accurate, but
they do not need to be as long as we are dealing only with
the apparent diameter of the moon, which changes slowly with
time.

Tycho gave the time by means of a 12-hour clock, which
I have translated into a 24-hour clock reading, measured
from midnight. When Tycho gave the time of an observation
to a precision of more than a minute, he usually did so by
stating some simple fraction of a minute, which I have
changed into seconds. For example, on 1585 February 28,
Tycho said that the time was 7 hours, 25 1/3 minutes in the
evening. I have translated this into 19:25:20 hours.

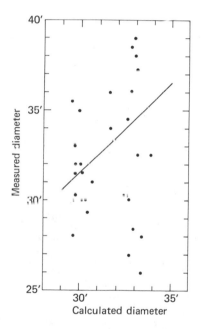

Figure II.1 The measured diameter of the moon plotted against its calculated diameter,
from Tycho's data. On the average, the measured diameter is 1'.52 greater than the
calculated value. The three points closest to the lower right-hand corner were measured
in the daytime and are not included in this average. The straight line is the calculated
value plus 1'.52.

---

†Tycho, as a citizen of Denmark, used the Julian calendar
until he left Hveen and joined the court of the Holy Roman
emperor. The dates in Table II.1 are in the Julian calendar, as all other dates of observations in this volume will
be.

Tycho did not give the time of seven of these observations. I have estimated the times of these observations by considering adjacent observations for which he did record the time. These estimated times are labelled as such in the table.

The third column gives the diameter of the moon that Tycho stated. When he stated the diameter to a precision better than 1', he usually did so by means of a simple fraction, which I have converted to decimal form. The reader can readily reconstruct the fraction that Tycho stated, if he should want it.

The fourth column gives the diameter of the moon calculated from modern theory. In the calculations, I assume that the times listed in the second column give local mean time at Hveen, and that Hveen is $0^h.846\,574$ east of Greenwich. To find the diameter, I first calculate the distance ρ_M from the observer (not from the center of the earth) to the center of the moon. I then take the diameter to be $1\,873'.45/\rho_M$.

The significance of the last column will be explained in a moment.

Let M denote a measured value in Table II.1 and let C denote the corresponding calculated value. M is plotted as a function of C in Figure II.1. The scatter of the points is so great that it is not obvious that there is any significant relation between M and C. Actually, things are not as bad as they look from the figure, as we find out if we go ahead and find the straight line that gives the best fit to the points. Before we do this, however, we should note one other matter.

The three observations made on 1585 September 8 and 10 were made during daylight. During the daytime, the moon is but little brighter than the sky, and the effects which make the moon appear too large at night should be absent during the day. If so, the daytime observations should not be mixed with the other ones, and Figure II.1 confirms this. The three points in question are the ones that are closest to the lower right-hand corner of the figure, and I think we are justified in omitting these points from the analysis.

Now we want to find the straight line that gives the best fit to the points in Figure II.1, after we omit the daytime observations. In finding this line, we should refer the values of C to their mean value, which is $31'.32$, rather than to $C = 0$. When we do this, we find

$$M = C + (1'.52 \pm 0'.57)$$
$$+ (-0.04 \pm 0.40)(C - 31'.32). \tag{II.2}$$

That is, the derivative dM/dC is 0.96, which is not significantly different from unity.

We are entitled to conclude that Tycho followed the
variations in the diameter of the moon correctly, but that
he measured the diameter with a bias and with a random
error. If this is so, then M should equal C + A, in which A
is a constant, plus a random error. When we evaluate A, we
find:

$$\underline{M} = \underline{C} + 1'.52 \pm 0'.57. \qquad\qquad (II.3)$$

The relation M = C + 1'.52 is the straight line in Figure
II.1, and the residuals M − C − 1'.52 are tabulated in the
last column of Table II.1.

The standard deviation of the residuals is 2'.90, but
this is the resultant of two independent measurements of a
limb position. The standard deviation of a single observation is 2'.90/√2, which we can take as exactly 2' for simplicity.

From Equation II.3 we find that Tycho did, in fact,
overestimate the size of the moon by 1'.52, so he presumably
had a bias in locating the position of a single limb equal
to 0'.76, about 46". If we knew the size of this bias with
enough accuracy, we could simply correct his lunar data by
this amount. However, the uncertainty in the bias is 0'.57,
about 34", for the diameter and about 17" for a single limb.
This far exceeds our allowance of 4" for an unknown bias,
and thus we cannot simply correct Tycho's observations of
the azimuth of a single limb.

Instead, we must analyze measurements of the eastern
limb and of the western limb separately, and we must then
combine them by assigning equal weights to each body of
data, whether or not they contain equal numbers of observations. This would not be a serious limitation if measurements of the eastern and western limbs both occurred in
sufficiently large numbers, but they do not. An overwhelming fraction of the observations are of the western limb.
Of the first 66 observations that I counted, 62 were of the
western limb and only 4 were of the eastern limb. Tycho
seems to have made only about 1 observation per year of the
eastern limb.† Altogether, I think we will be optimistic if
we assume that there are 100 suitable observations of the
eastern limb. If each is subject to a random error of 2' or
120", the resultant error is still 12". It would take 900
observations to reduce the effect to 4", which is our thresh-
old.

Our conclusion is that the observations in which Tycho
measured the position of the moon and of a star or planet at
almost the same time will not make a useful contribution to
our knowledge of the earth's acceleration.

---

†This remark is confined to observations for which Tycho
 also measured the position of a star or planet almost
 simultaneously.

5. Simultaneous Measurements of Azimuth and Elevation

We now turn to the other type of observation that holds an initial promise of being useful. In this type, Tycho measured the azimuth of either the east or west limb (usually the west) at the same time that he measured the elevation of either the upper or lower limb.† Luckily, measurements of the upper and lower limbs occur with nearly equal frequency, so we will not encounter the extreme imbalance that we found in the preceding section.

In order to use such an observation, we use the azimuth measurement to find the time, assuming a value for the lunar acceleration in doing so. The uncertainty in the lunar position that results from the present uncertainty in the acceleration is about 4″, as we have seen. This much error in a lunar position used to find the time creates a time error of only 4/15 seconds, and the moon moves only 2/15 seconds of arc during this time. This is indeed negligible.

In fact, we do not even have to worry about the uncertainty in where Tycho placed the eastern or western limb. We see from the discussion below Equation II.3 that this uncertainty is about 17″, and this causes an error of about $0^S.5$ in inferring the time.

Since the observations of the upper and lower limbs occur in approximately even numbers, we can reduce the random error from 120″ to 4″ by using 900 observations, as we saw in the preceding section. I have not counted the total number of observations of this sort, but I did count several hundred within a span of 4 years, so it is likely that there are more than 900 observations. If so, our worry is not about the random errors present in the observations, but about the systematic ones.

We can study the possibility of systematic errors in the readings of elevation by turning to Tycho's observations of the meridian elevation of the sun and stars. I will analyze some of these in the next section.

6. Meridian Elevations of the Sun and Stars

Tycho made thousands of measurements of the meridian elevation of the sun, the moon, and the stars. In this section, we want to study the possibility of systematic errors in these measurements, the extent to which we can measure the systematic errors by means of the stars, and the use of these results to correct measurements of the sun and moon. In order to do this study, we need to use observations made within the same general time span and on the same instrument.

---

†Sometimes he measured the right ascension and declination rather than azimuth and elevation. The principles of analysis are the same either way, and I will speak as if all the measurements were of azimuth and elevation.

After he acquired several observing instruments, Tycho developed a habit of using instruments for observing the sun which were usually different from those used with the stars. We need observations that were made before he developed this habit. So far as I noted in going through the record of his observations, the most useful observations for our purposes are those made with the great mural quadrant in 1583. The great mural quadrant† was a quadrant with a radius slightly more than 2 meters, which was graduated by intervals of 10″. It had a fixed sight at its center and a movable sight that could slide along the graduated arc. The quadrant was fixed to the interior of the west wall of the observatory, which was carefully oriented to lie in the meridian plane. Thus, the quadrant was shielded from sunlight except for a few minutes at noon.‡

The mural quadrant was at latitude 55° 54′ 24″.67 and at longitude 12° 41′ 54″.98 [Dreyer, 1926, volume 10, p. XXVI]. I have not found a precise statement of its altitude above sea level, but Dreyer [1890, p. 93] says that it was about 50 meters. Luckily, we do not need to know the altitude accurately. The latitude and longitude were measured during a careful survey of the site made in 1903 - 1904.

Before we can use the elevation measurements of the sun and stars, we must correct them for refraction and, in the case of the sun, for parallax. The effect of aberration on a meridian elevation of the sun is small and I will ignore it. Although Tycho actually measured the elevation, it is easier to make the corrections if we use the zenith distance, and I will change Tycho's measurements into the equivalent zenith distances in the rest of this chapter. I will use ζ to denote the measured (apparent) zenith distance and $\underline{Z}$ to denote the corrected value.

In order to correct for solar parallax, we subtract a quantity Π from the apparent zenith distance ζ, where

$$\Pi = 8″.80 \sin \zeta. \qquad (II.4)$$

In order to correct both the solar and stellar distances for refraction, we add a quantity R to ζ. It is easiest to break the calculation of R into parts. First, we calculate R at the conventional surface atmospheric pressure

---

†Almost all works about Tycho, even rather short ones, show the picture of this instrument that Tycho drew and published himself [Brahe, 1598, folio B,1]. I do not think it is necessary to print the picture again here. Dreyer [1890] reproduces it following his page 100, and Wesley [1978] prints it on his page 46.

‡I do not know Tycho's reasons for this arrangement, but I doubt that it was to minimize heating effects on the instrument. I imagine that it was done for comfort and convenience. Also, while the instrument itself was shielded from the sun, the wall to which it was attached was exposed directly to the sun during the entire afternoon.

of 760 millimeters of mercury and the conventional surface
temperature of 10° centigrade. _Smart_ [1962, Section 37]
shows that R under these conditions can be expanded in the
form:

$$R = \underline{A} \tan \zeta + \underline{B} \tan^3 \zeta.$$

In principle, it is possible to calculate $\underline{A}$ and $\underline{B}$ from
the molecular properties of the atmosphere, but Smart points
out that it is safer to calculate them by fitting to measured values of R as a function of ζ. I have done so by
using the table of atmospheric refraction given by _Allen_
[1962, p. 120], with the results that $\underline{A}$ = 58″.294 and $\underline{B}$ =
−0″.0504.

When the surface temperature is θ°centigrade and the
surface pressure is $\underline{p}$ millimeters of mercury, we multiply
the conventional value of R by the factor $\underline{p}$/[760(0.962
+ 0.0038θ)] [Allen]. When we include this factor, and use
the values of $\underline{A}$ and $\underline{B}$ already found, we can write R in the
form:

$$R = (58″.294 \tan \zeta - 0″.0504 \tan^3 \zeta)(\underline{p}/760)$$

$$\div (0.962 + 0.0038θ). \qquad\qquad (II.5)$$

Equation II.5 represents Allen's table with a maximum error
of about 0″.6 for values of ζ between 0° and 75°, but it
begins to diverge soon after. For safety, I will not use
any value of ζ greater than 70°.

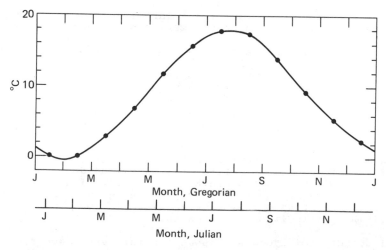

Figure II.2 **The mean temperature as a function of the season at a station near Hveen.**
The station, called Landbohojskolen/Kobenhavn, Denmark, is about 25 kilometers from
Hveen. The beginnings of the months are indicated by the tick marks in (a) the Gregorian
calendar, which we may take as independent of the year, and (b) the Julian calendar in
the time of Tycho.

Since the altitude of Tycho's observatory was about 50 meters, we can take p = 756 millimeters, which is the mean pressure at this altitude, for all observations. In order to estimate the temperature Θ at the time of each observation, I turn to the data supplied by the National Oceanic and Atmospheric Administration [1976]. This publication gives, month by month, the normal mean temperature for many places on earth. The nearest place to Hveen I can find is a station called Landbohojskolen/Kobenhavn, which is about 25 kilometers from Hveen.

Figure II.2 shows the monthly means for this station plotted at the midpoints of the respective months. A smooth curve joining the points allows us to make a good estimate of the normal mean for any day in the year. Since the months used in gathering the data are in the Gregorian calendar, it is sufficiently accurate to let the figure apply to any year, provided that we use the Gregorian calendar. However, Tycho used the Julian calendar. The lower set of the tick marks in Figure II.2 shows the beginnings of the various months in the Julian calendar in Tycho's time.

There is one further complication in using the temperature. The temperature plotted in Figure II.2 is the mean over both the day of the month and the hour of the day. The star observations tended to be made between sunset and midnight, and the temperature when they were made was probably close to the mean for the day. Thus, for a stellar observation, we simply read off the temperature for a particular date from the figure. The meridian zenith distance of the sun, however, was necessarily measured at noon, when the temperature was usually above the average. As an approximation, I add 3° to the temperature shown in the figure for a solar observation.

TABLE II.2

ERRORS IN TYCHO'S MEASUREMENTS OF MERIDIAN ZENITH
DISTANCES OF STARS MADE IN 1583

Date of first observation in a group	Date of last observation in a group	Average error[a] "	Number of observations
Jan 18	Jan 21	8.7	17
Jan 22	Jan 27	35.2	19
Feb 26	Apr 3	23.1	8
Apr 19	May 3	25.8	11
Aug 5	Aug 14	46.4	14
Aug 16	Aug 27	50.9	15
Sep 4	Sep 16	47.4	12
Sep 23	Sep 30	44.1	12
Oct 10	Oct 30	38.9	7
Dec 2	Dec 27	32.4	8

[a]Measured value minus calculated value

Now we can turn to the stellar observations that Tycho made with the mural quadrant in 1583. In his notes, Tycho expressed doubt about the quality of some of the observations, marking them with non satis bona or some equivalent phrase; I omit these observations from the analysis. I also omit those for which the zenith distance was greater than 70°, for a reason already stated. This leaves us with 123 observations.

These observations are summarized in Table II.2. In preparing this table, I divided the observations into ten groups, chosen to have approximately the same number of observations in each group without covering too long a time span in any single group. In the table, the first column gives the date of the first observation in a group and the second gives the date of the last observation in a group. The third column gives the average error of the measured zenith distances for that group, by which I mean the measured value minus the value taken from modern tables and rotated back to the date of the observation (including the proper motions, the precession, and the rotation of the ecliptic). The last column gives the number of observations in each group.

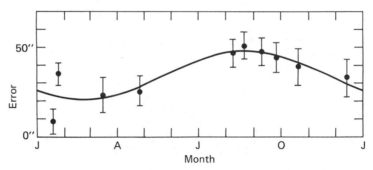

Figure II.3 The seasonal dependence of Tycho's star observations in 1583. The points are the average errors, in the sense of measured values minus correct ones, for groups of meridian zenith distances measured within a short time interval; an error bar is associated with each average. The curve is the sum of a constant plus a sinusoid that best fits the errors.

The reader can probably detect the seasonal dependence of the errors from the table, but Figure II.3 makes it quite obvious. The figure shows each average error in Table II.2, along with an associated error bar, plotted against the average date of a group of observations. If there is a seasonal effect, we expect the error to be a constant plus a function of time that is approximately sinusoidal. The sinusoidal function $\underline{z}$ that gives the best fit to the points in the figure is

$$\underline{z} = 34''.3 \pm 3''.1$$

$$- (13''.4 \pm 5''.1) \sin (n_{\underline{s}}\underline{D} + 38°.8 \pm 5°.9). \qquad (II.6)$$

In this, $\underline{D}$ is not the lunar elongation as it usually is, but rather the day of the year, with $\underline{D} = 0$ on January 1. $n_{\underline{s}}$ is the mean motion of the sun.

The function $34''.3 - 13''.4 \sin (n_{\underline{s}}\underline{D} + 38°.8)$ is plotted in Figure II.3. The standard deviation of the residuals from this function is only $26''.4$. That is, this is the precision of Tycho's stellar observations made with the mural quadrant. This is far better than the precision we found for his observations of a limb of the moon. It is also the basis for calculating the lengths of the error bars in Figure II.3.

Now let us turn to the measurements of the zenith distance of the sun that Tycho made in 1583. These are listed in Table II.3. The first column in the table gives the date of an observation and the second gives the result of the observation. Tycho actually measured the elevation angle in degrees, minutes, and simple fractions thereof.[†] I have translated these data into zenith distance given in degrees and decimal fractions.

I do not use all the observations in the analysis. I delete those for which the zenith distance is greater than 70°, for a reason that has already been explained, and I also delete the ones which Tycho specifically said were of doubtful quality. This leaves 94 observations that will be used. The third column gives the value of the zenith distance for these 94 observations that I calculate taking the value of $\underline{y}$ to be zero. I will let Z_o denote a value calculated this way. I do not believe it is necessary to give the details of these calculations except to remind the reader that the time of the observation is apparent noon rather than mean noon at Hveen.[‡]

[†]Tycho gave the number of seconds a few times.

[‡]I did the calculations for Table II.3 before I received the tapes which contain the DE102 ephemeris. Thus, I did the calculations using Newcomb's theory [Newcomb, 1895] rather than the new ephemeris. If I were going to use the observations in Table II.3 in an inference of $\underline{y}$, I would repeat the calculations using the new ephemeris. However, for reasons that we will see in a moment, I will not use the observations in the table, and thus I do not think it is necessary to repeat the calculations. The error in Newcomb's theory is much less than the systematic error in Tycho's observations.

TABLE II.3

VALUES OF THE MERIDIAN ZENITH DISTANCE OF THE SUN MEASURED IN 1583

Date	Measured with the mural quadrant °	Calculated with y=0 °	$10^5 \times$ dZ/dy	Residual "
	Zenith distance			
Jan 18	74.1458			
Jan 19	73.8944			
Jan 21	73.3500			
Jan 26	71.9000			
Jan 27	71.5972			
Feb 2	69.7500			
Feb 7	67.9833	68.0031	+4.56	+32
Feb 22	62.4667	62.4737	+5.02	+45
Mar 2	59.3500	59.3648	+5.14	+ 4
Mar 9	56.6083	56.6069	+5.16	+52
Mar 10	56.2056	56.2122	+5.16	+22
Mar 13	55.0250	55.0295	+5.14	+26
Mar 14	54.6292			
Mar 24	50.7500	50.7484	+5.00	+36
Mar 26	49.9833	49.9862	+4.95	+18
Mar 27	49.6083	49.6076	+4.94	+30
Mar 28	49.2222	49.2308	+4.91	- 5
Mar 31	48.1194	48.1120	+4.84	+50
Apr 2	47.3750			
Apr 4	46.6500			
Apr 5	46.3000			
Apr 8	45.2333	45.2295	+4.56	+31
Apr 9	44.8750	44.8812	+4.52	- 6
Apr 14	43.1806	43.1863	+4.32	- 8
Apr 15	42.8500	42.8574	+4.26	-15
Apr 16	42.5333	42.5320	+4.22	+16
Apr 18	41.8875	41.8922	+4.13	- 7
Apr 19	41.5750	41.5780	+4.08	- 2
Apr 20	41.2583	41.2678	+4.02	-25
Apr 21	40.9583	40.9615	+3.98	- 3
Apr 22	40.6500	40.6594	+3.91	-26
Apr 23	40.3542	40.3614	+3.86	-19
Apr 24	40.0611	40.0676	+3.81	-17
Apr 25	39.7667	39.7782	+3.76	-36
Apr 26	39.4861	39.4932	+3.70	-20

TABLE II.3 (continued)

Date	Zenith distance		$10^5 \times$	Residual
	Measured with the mural quadrant °	Calculated with $\underline{y}=0$ °	$d\underline{Z}/d\underline{y}$	"
Apr 28	38.9333			
May 1	38.1333	38.1371	+3.39	−11
May 5	37.1375	37.1399	+3.12	− 8
May 6	36.9083	36.9035	+3.06	+18
May 7	36.6750	36.6724	+2.98	+10
May 8	36.4528	36.4466	+2.91	+22
May 10	36.0111	36.0116	+2.77	− 3
May 11	35.8000	35.8025	+2.69	−10
May 12	35.5944	35.5990	+2.63	−18
May 13	35.4000	35.4013	+2.55	− 7
May 14	35.2056	35.2094	+2.47	−17
May 15	35.0222	35.0234	+2.39	− 7
May 16	34.8431	34.8434	+2.31	− 4
May 17	34.6681	34.6694	+2.22	− 8
May 20	34.1917	34.1840	+1.99	+23
May 21	34.0403	34.0347	+1.91	+15
May 22	33.9000	33.8916	+1.83	+25
May 25	33.4958	33.5008	+1.57	−24
May 29	33.0611	33.0706	+1.24	−41
Jun 3	32.6833	32.6826	+0.79	− 6
Jun 5	32.5792			
Jun 6	32.5333	32.5313	+0.53	− 1
Jun 9	32.4389			
Jun 11	32.4167	32.4164	+0.08	− 9
Jun 14	32.4306			
Jun 16	32.4722	32.4737	−0.36	−15
Jun 17	32.5000	32.5058	−0.47	−31
Jun 18	32.5417	32.5447	−0.56	−21
Jun 19	32.5833	32.5905	−0.64	−36
Jun 20	32.6347	32.6430	−0.73	−40
Jun 21	32.6917	32.7024	−0.82	−49
Jun 22	32.7611	32.7685	−0.91	−37
Jun 23	32.8333	32.8413	−1.00	−39
Jun 26	33.0917	33.1000	−1.26	−40
Jun 30	33.5278	33.5372	−1.59	−44
Jul 2	33.7917	33.7947	−1.77	−21
Jul 5	34.2306			
Jul 6	34.3778			
Jul 11	35.2833			
Jul 12	35.4611			

TABLE II.3 (continued)

| Date | Zenith distance | | $10^5 \times$ | Residual |
	Measured with the mural quadrant °	Calculated with $\underline{y}=0$ °	$d\underline{Z}/d\underline{y}$	"
Jul 13	35.6500			
Jul 15	36.0750	36.0674	-2.77	+19
Jul 23	37.9375			
Jul 28	39.2611	39.2692	-3.63	-34
Jul 29	39.5417			
Aug 3	41.0167			
Aug 4	41.3167			
Aug 6	41.9444	41.9432	-4.11	+ 4
Aug 10	43.2250	43.2326	-4.29	-26
Aug 11	43.5542	43.5635	-4.34	-32
Aug 12	43.8875	43.8977	-4.38	-35
Aug 13	44.2278			
Aug 14	44.5750	44.5756	-4.45	+ 1
Aug 15	44.9167	44.9191	-4.48	- 5
Aug 16	45.2667	45.2654	-4.54	+ 9
Aug 19	46.3167	46.3211	-4.64	- 9
Aug 23	47.7667	47.7639	-4.77	+19
Aug 24	48.1333	48.1304	-4.78	+21
Aug 25	48.5000	48.4990	-4.82	+15
Aug 27	49.2417	49.2422	-4.88	+10
Aug 28	49.6167	49.6167	-4.89	+13
Aug 30	50.3694	50.3710	-4.93	+ 9
Sep 1	51.1333	51.1317	-4.97	+23
Sep 2	51.5167	51.5142	-4.99	+27
Sep 4	52.2806	52.2830	-5.02	+11
Sep 5	52.6667	52.6691	-5.03	+12
Sep 8	53.8333	53.8331	-5.07	+25
Sep 11	54.9944	55.0035	-5.08	- 5
Sep 12	55.4000			
Sep 13	55.7806	55.7861	-5.10	+10
Sep 14	56.1667	56.1778	-5.11	- 9
Sep 16	56.9556	56.9614	-5.10	+12
Sep 23	59.6917	59.6974	-5.07	+23
Sep 24	60.0764	60.0862	-5.06	+10
Sep 28	61.6111			
Sep 29	62.0028	62.0162	-4.98	+ 5
Sep 30	62.3833	62.3986	-4.98	0
Oct 1	62.7667	62.7797	-4.95	+10
Oct 10	66.1083	66.1265	-4.71	+10
Oct 12	66.8306	66.8451	-4.64	+28

TABLE II.3 (continued)

Date	Zenith distance		$10^5 \times$	Residual
	Measured with the mural quadrant	Calculated with $\underline{y}=0$	$d\underline{Z}/d\underline{y}$	
	°	°		"
Oct 14	67.5333	67.5528	-4.56	+15
Oct 15	67.8792	67.9021	-4.54	+ 5
Oct 16	68.2306	68.2484	-4.48	+26
Oct 20	69.5806	69.6004	-4.30	+30
Oct 21	69.9056	69.9295	-4.26	+17
Oct 24	70.8708			
Oct 26	71.5000			
Oct 30	72.6792			
Nov 7	74.8333			
Nov 8	75.0625			
Dec 2	78.9514			
Dec 3	79.0278			
Dec 11	79.3389			
Dec 14	79.3167			

     I omit the calculated value for those observations that
will not be used in the analysis. As it happened, Tycho did
not mark any observation "doubtful" for which the zenith
distance was greater than 70°. Thus the reader can tell
whether a particular observation is being omitted because it
was of doubtful quality or because the sun was too far from
the zenith.

     The fourth column in Table II.3 gives the derivative of
the calculated zenith distance $\underline{Z}$ with respect to $\underline{y}$. I
calculated this derivative numerically in the obvious way.

     Let $\underline{Z}_a$ denote a measured value of the zenith distance,
after it is corrected for parallax and refraction in the
manner described at the beginning of this section. $\underline{Z}_a$ is
subject to random errors of observation, to a bias $\underline{M}$ in
Tycho's establishment of the vertical, and to other systematic errors that we will ignore for the moment. If we
ignore them, then we want to find the constants $\underline{M}$ and $\underline{y}$
which make the function

$$\underline{M} + \underline{Z}_0 + \underline{y}(d\underline{Z}/d\underline{y})$$

a best fit to the $\underline{Z}_a$. This is a standard procedure that
should not need description. The results are:

$$\underline{M} = 40''.5 \pm 2''.4, \qquad \underline{y} = -32.0 \pm 16.9. \qquad (II.7)$$

These values tell us two things.

First, the standard deviation attached to $\underline{y}$ is only
16.9 for a single year's observations. With 20 years of
observations, this would fall to about 4 if there are no
important systematic errors other than $\underline{M}$, and the observations would make a useful contribution to finding $\underline{y}$. Second, the bias $\underline{M}$ in Equations II.7 does not differ significantly from the bias that we found from the stellar observations in Equation II.6. This bias presumably arises from
Tycho's error in finding the local vertical. It would be
interesting to know the result of a modern measurement of
the deflection of the vertical on Hveen.

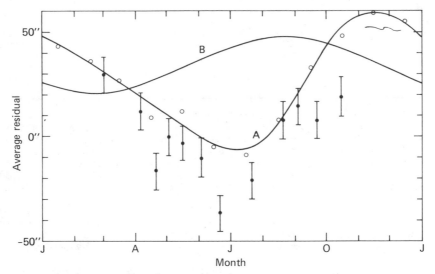

Figure II.4 The seasonal dependence of Tycho's solar and stellar observations. Curve B
shows the dependence of the stellar observations in 1583; it is the same as the curve in
Figure II.3. The open circles are the average errors in the solar zenith distances for the
years 1587 − 1590 [Tupman, 1900, p. 169], and curve A is a smooth curve I have drawn
using these points. The solid circles, with associated error bars, are the errors I find in
1583 for various groups of observations.

The last column in Table II.3 gives the residual of
each observation, which means the measured value minus the
fitted function. The standard deviation of the residuals is
$23''.1$. This is almost identical with the value $26''.4$ that

we found for the stellar observations. This confirms that
Tycho's observations of a limb of the moon are considerably
less precise than his observations of the sun or stars. His
precision with the sun and stars (and presumably the planets) was between 20" and 30", but his precision with a limb
of the moon was about 2' (Section II.4).

I said above that Tycho's observations of the meridian
elevation of the sun would make a valuable contribution to
our work if there were no systematic error other than the
bias M. Unfortunately, the errors in the solar observations,
like those in the stellar observations, have a seasonal
dependence. We can see this easily from Figure II.4. For
this figure, I have gathered the residuals into twelve
groups, with seven residuals in the sixth and twelfth groups
and eight in the others; for each group, I have then found
the (algebraic) average residual and the standard deviation
of the average. The averages and their standard deviations
appear as the solid circles and associated error bars in
Figure II.4. It is clear that the errors are not random.

We found earlier that the errors in the meridian elevations of the stars were not random, and the curve marked B
in Figure II.4, which is the same as the curve in Figure
II.3, shows the systematic error in the stellar observations. Curve B includes the bias in elevation, which has
been removed from the plotted points. Even when we correct
for the difference in bias, I think it is clear that curve D
does not represent the error in the solar observations at
all well. Curve B is almost exactly 180° out of phase with
the solar errors.

The difference between the solid circles and curve B is
not a sampling accident, as we see from the work of Tupman
[1900, p. 169]. Tupman found the average error in Tycho's
measurements of the meridian zenith distance of the sun for
each month during the years 1587 - 1590, as well as the
average over these four years for each calendar month. I
have plotted Tupman's averages as the open circles in Figure
II.4, and curve A is a smooth curve that I drew by eye
through the open circles before I plotted the solid circles
and error bars.

Curve A includes the bias in the measurements, which
has been removed from the solid circles. If we drop curve A
by 30" or so, we see that it agrees quite well with the
solid circles. This is significant, because the curve was
derived from the years 1587 to 1590 while the points were
derived from the year 1583 alone. In other words, while the
mural quadrant had some property that changed by a large
amount with the seasons, its properties at a given season
did not change appreciably from year to year.

Since curves A and B, which apply to the solar and
stellar observations, respectively, are almost out of phase,
the seasonal effect cannot be simply a tilting of the entire
instrument due to a change in temperature. The causes of
the seasonal bias must have been quite complex.

Because of the seasonal effect, we cannot take the solution for y in Equations II.7 to be valid. Since the derivative dZ̄/dy necessarily varies with a period of a year, any systematic error with a period of a year directly "resonates" with the meridian position of the sun when we solve for y.† To get some idea of the sensitivity of y to a seasonal error in the observations, let us assume that curve B does represent correctly the seasonal effect in the solar data and make a new determination of the bias M̲ and the acceleration y̲. The results are:

$$\underline{M} = 38''.6 \pm 2''.5, \qquad \underline{y} = +31.2 \pm 17.8. \qquad (II.8)$$

We hardly change the value of M̲; this is a consequence of the fact that curves A and B in Figure II.4 show almost the same bias. The value of y̲, however, changes by a large amount, by 63.2. Since the amplitude of the oscillation in curve B is only 13''.4, this means that a seasonal oscillation with an amplitude of only 1'' can change the inferred value of y̲ by 4.7.

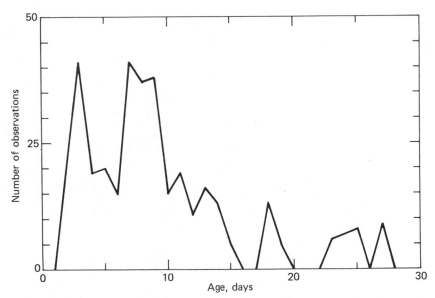

Figure II.5 The age of the moon at the time of Tycho's measurements of the lunar altitude. The horizontal axis is the age of the moon, averaged to the nearest day, and the vertical axis is the number of measurements made when the moon had that age. The figure is based upon the measurements made in the years 1578 to 1585.

†This is not quite true. See Section III.3 for a yearly effect that can be removed.

When we compare Equations II.7 and II.8, we see that
the quality of the fit goes down slightly when we use curve
B. Further, the standard deviation of the residuals rises
from 23".1 to 24".4. These facts give further evidence that
the stellar and solar observations have a different seasonal
dependence.

If the stellar and solar observations had the same
seasonal dependence, we could safely apply this dependence
to the measurements of the lunar elevation and thence find
the acceleration of the moon (with respect to solar time).
The fact that the seasonal dependence is different for the
different classes of observation means that we cannot proceed in this simple way. To be sure, we might try to use
the following argument: The difference between the effects
on the stellar and solar observations probably comes from
the fact that the solar observations were made at noon and
the stellar ones at night, mostly between sunset and midnight. (This would mean that the difference in seasonal
dependence is probably a temperature effect.) Since the
lunar observations tended to be made at the same time of day
as the stellar ones, the stellar dependence is the one we
should use in correcting the lunar observations.

If we could be sure that the seasonal error was, in
fact, a function of just the season and the time of day,
this argument would be valid. However, I do not feel that
we can use this argument with confidence, because the seasonal error is clearly a complex matter. If it did result
simply from the temperature at the time of an observation,
the curves A and B in Figure II.4 would have the same phase,
but we see that they are almost exactly out of phase.

Thus we conclude that we cannot remove the seasonal
effect from the lunar data. This would not affect the inference of the lunar acceleration, however, if the lunar
observations were well distributed over all phases of the
moon. Unfortunately, Tycho's measurements of the lunar
elevation are not well distributed, as Figure II.5 shows
us. In drawing the figure, I took the simultaneous measurements of azimuth and elevation made during the years from
1578 to 1585; I counted 367 measurements during this period.
For each measurement, I calculated the age of the moon,
rounded to the nearest day, and I plotted the number of
measurements made at each possible age, which ranges from 0
days to 30 days.

We see that the measurements are strongly biased toward
the first half of the month, and that they show a sharp peak
near the first quarter. Tycho made more than six times as
many measurements during the first half of a lunar month as
he did during the second half. In order to obtain a sample
that was unbiased with respect to the age or phase of the
moon, we would have to do the statistical equivalent of
throwing away more than two-thirds of the measurements, and
this would not leave us with enough measurements to make a
useful estimate of the lunar acceleration.

The distribution of the age of the moon in Figure II.5
is a consequence of the fact that Tycho kept hours that seem

normal to most Americans I know. That is, he usually stayed
up for a number of hours after sunset, but he got up before
sunrise only for exceptional occasions. This affects the
age of the moon when he saw it. Most of the time when the
moon rises in the evening, it is between the new and full
moon, so observing in the evening biases the age of the moon
heavily toward the first half of the lunar month.

I regretfully conclude that we cannot use Tycho's observations, in spite of their high quality and large number,
in studying the accelerations of the moon and of the earth's
spin.

7.   Nicolaus Copernicus

It is pointless to ask for the correct spelling of this
astronomer's name, since he lived in an age when people were
not concerned with consistency in spelling, particularly of
proper names. He is generally known by the Latinized form
of his name, which is probably spelled Nicolaus Copernicus
more often than any other way.

Copernicus was born in Torun on 1473 February 19.†
Torun was a Polish city at the time, and it has passed between Polish and German (or Prussian) possession more than
once since then. It has been Polish since 1919. Copernicus's father was a successful merchant who married into a
wealthy family named Waczenrode. Thus Copernicus, like
Tycho, had the advantage of family wealth and position behind him.

Copernicus's father died when he was ten years old, and
his maternal uncle, Lucas Waczenrode, adopted him and became
responsible for his education. Lucas was a canon of the
cathedral of Frombork (German Frauenburg) at the time, and
he became the bishop six years later. About 1491 or 1492,
Copernicus was sent to the University of Krakow, where he
became interested in mathematics and astronomy, perhaps for
the first time. About 1496, he was sent to Italy for fur-

---

†The 500th anniversary of Copernicus's birth occurred in
1973, and it prompted a large outpouring of publications
about him. Most of these are highly specialized and I did
not use them in the preparation of this section. Aside
from Copernicus's own writing, this section is based almost
entirely upon Rosen [1971] and Armitage [1951]; the former
is an article in the Dictionary of Scientific Biography and
the latter is a "popularized" book. Rosen [p. 411] writes:
"The standard biography by Leopold Prowe, Nicolaus Copper-
nicus, 2 vols. (Berlin, 1883-1884; repr. Osnabrück, Ger-
many, 1967), has not yet been superseded, even though it
is incomplete (the planned third volume was never published), nationalistically biased, scientifically inade-
quate, somewhat inaccurate, and partly obsolete." (The
parentheses, abbreviations, and so on are in the original).
I did not consult this work in spite of Rosen's praise
of it. Rosen does not refer to Armitage's biography.

ther education. He spent about ten years in Italy as a
student and, apparently, as a lecturer. He studied canon
law and medicine, and it seems that he lectured on mathematics and astronomy at Bologna, Rome, Padua, and Ferrara.
He was installed as a canon of the cathedral of Frombork on
1501 July 27,† and his first act after his installation was
to apply for leave of absence so that he could continue his
education in Italy. This leave was granted.

Copernicus finally returned from Italy in 1506 with a
truly remarkable education. He was now 33 years old and he
had been a student almost all his life. He had received an
advanced education in astronomy, mathematics, canon law, and
medicine. Even so, he still did not take up his responsi-
bilities as a canon. Instead, his uncle, the bishop, appointed Copernicus as his personal physician, and this required that Copernicus receive a further leave of absence
from his position.‡ Not surprisingly, this leave was also
granted.

When his uncle died in 1512, Copernicus finally went to
Frombork to take up his duties as canon, about 15 years
after his election and 11 years after his formal installation. He remained in the service of the cathedral chapter
of canons until his death in 1543.

Shortly after he moved to Frombork, he built a small
observatory, but we may doubt that he made a systematic set
of observations. It is more likely that he made observations as he needed them. He published 27 observations in
his main book, and he noted still others in the margins of
books in his library, where they can still be seen [Armi-
tage, 1951, p. 85].

His main work is De Revolutionibus Orbium Caelestium
Libri Sex [Copernicus, 1543], which is usually called De
Revolutionibus for short. The latest edition of De Revolutionibus that I know of is the one by Kubach that is mentioned in the list of references. There is an English
translation in volume XVI of the series called Great Books
of the Western World.‡ By a series of happenstances, a
Lutheran clergyman named Andrew Osiander actually oversaw

---

†At the time, a canon of a major cathedral was automatically
a person of considerable wealth and influence. Copernicus
was actually elected as a canon about 1496 or 1497, but he
was not installed until 1501.

‡The bishop of Frombork did not live near the cathedral but
in the castle at Heilsberg, about 60 kilometers away.

‡I should warn the reader that the accuracy of this translation is poor, at least as far as the texts of the observations are concerned; I have not read many other parts of
the text. The reader who wants to know the facts of the
observations, and the use that Copernicus made of them,
should go to some other version or edition of the work. I
used the edition by Kubach that was just mentioned.

the publication of De Revolutionibus, and I believe it is now generally accepted by scholars that the introduction was written by Osiander and not by Copernicus.

Toward the end of 1542, Copernicus suffered an apoplexy which left him partly paralyzed, and he lived barely long enough to see a copy of De Revolutionibus. An advance copy reached him on 1543 May 24, and he died later that same day.† He was buried in the cathedral, but knowledge of the exact location of his grave has been lost.

Copernicus uses four main sources of data in determining the parameters that occur in his models of motion of the solar system. These are Ptolemy [ca. 142], some Arabic writing such as that of al-Battani [ca. 925], observations that he made himself, and three observations made at Nürnberg in his own time but not by himself. Medieval astronomers, both Arabic and European, could not reach agreement on the accuracy of the observations that Ptolemy claimed to have made but which, in fact, he fabricated [Newton, 1977]. Some astronomers, while accepting Ptolemy's general methods and theories, believed that his observations were inaccurate and did not use them. Others believed that his observations were accurate and did use them.

Unfortunately, Copernicus is in the second group. He accepted all of Ptolemy's observations, even his egregiously bad ones of the obliquity and of the equinoxes and solstices. Thus to Copernicus, the equinoxes, the obliquity, and the length of the tropical year behaved in complicated ways. In order to accommodate these complications, he made his own system much more complex than it needed to be. In fact, Copernicus's models used about twice as many circles (or parameters) as Ptolemy's models [Neugebauer, 1957, p. 204], and this excess complexity certainly did not encourage the adoption of Copernicus's system. Further, the complications just mentioned led Copernicus to conclude that the equinox is not a good reference for measuring longitude, and he adopted the star γ Arietis instead. That is, he assigned 0° as the longitude of this star, by definition.‡ Still further, he was led to regard the sidereal year as a more useful quantity than the tropical year. Thus Ptolemy's fabrications were still harming astronomy even fourteen centuries later.

Copernicus's own observations in De Revolutionibus include two measurements of the position of the star α Virginis (Spica), the times of one autumnal equinox and of two vernal equinoxes, and the time when the sun was midway between the autumnal equinox and the winter solstice. Coper-

---

†This statement has the aura of romance rather than of fact, and I do not vouch for it. However, it does seem that the arrival of the book and Copernicus's death were separated by only a short time.

‡The longitude of γ Arietis from the equinox was about 26°.5 in Copernicus's time.

nicus also measured the times associated with five lunar
eclipses, and he made three direct measurements of the position of the moon. For each outer planet, he made three
measurements of the time when the planet was in opposition,
as well as its position at a fourth time. For Venus, he
made one measurement of its position, and he made no observations of Mercury at all. He explained the latter by saying [De Revolutionibus, Chapter V.30] "that the Nile does
not give out vapors as the Vistula does with us." For Mercury, he uses the three contemporaneous observations made
in Nürnberg that I just mentioned.†

We can easily investigate the accuracy of Copernicus's
observations by looking at his two observations of the vernal equinox. He found [Chapter III.16] that the vernal
equinox in 1515 came on March 12 at 4^h.30 and that the equinox the following year came on March 11 at 4^h.33 [Chapter
III.13]; both times are local.‡ He does not state where the
observations were made, but both were made several years
after he took up regular residence at Frombork. Further, he
states specifically that he measured the autumnal equinox of
1515 at Frombork. Thus, we may safely assume that he measured the two vernal equinoxes there also.

The difference between the two equinoctial times is
365^d 0^h 2^m, but the correct interval was 365^d 5^h 48^m 48^s.
The error is about 5^h.78. This error comes only from the
precision of Copernicus's observations; any bias in his set
ting of the local vertical would cancel on subtraction.
Since the declination of the sun changes by about 1' per hour
near an equinox, the combination of the errors in the meridian altitude of the sun in the 2 years is about 6'. If we
make the optimistic assumption that the errors were equal
but opposite, each error was about 3'.

Thus we conclude that Copernicus's precision in measuring the meridian altitude of the sun was about 3' or
worse. This should be the angle that he could measure with
the greatest precision, so all other errors, and particularly errors in longitude, should be much worse. Thus we
should assign negligible weights to Copernicus's observa-

---

†There are two oddities about this set of observations of
  Mercury. It will be more convenient to discuss them in
  Section III.2, which deals with Bernard Walther.

‡Copernicus actually states the times in a rather complicated fashion that I will not give.

tions when we are trying to find the astronomical accelerations.†

Although Copernicus was outstanding in his boldness of thought and in his theoretical ability, he was not outstanding as an observer, as many have pointed out before me. In fact, the accuracy of his observations is only about that which Hipparchus achieved 17 centuries earlier. It is considerably poorer than the accuracy that his contemporary, Bernard Walther, achieved, as we will see later, and it is far inferior to Tycho's accuracy fifty years later. We now turn to Bernard Walther and to his teacher Regiomontanus in Chapter III.

---

†To guard against the possibility that this large error comes from a recording error rather than from a genuine error of observation, I have also studied the timing error in some other observations. I find that it is of the order of 20 minutes, standard deviation. If we repeat the analysis found near the beginning of Section II.2, we find that we need the timing errors to be only about 20 seconds rather than 20 minutes.

THE LUNAR OBSERVATIONS OF REGIOMONTANUS AND WALTHER

1.   Regiomontanus and Walther

Although he was only 39 years old at the time of our
last word about him, Regiomontanus was probably the most
famous astronomer of the 15th century in Europe. His actual
name was perhaps the equivalent of John the Miller of King's
Mountain, but he became known to history by the Latinized
name of his birthplace. The longest account of his life and
work I have consulted is by Rosen [1975].

He was born on 1436 June 6 and he enrolled in the University of Vienna on 1450 April 14. He received his bache-
lor's degree on 1452 January 16, when he was not yet 16, but
the university regulations would not allow him to receive
his master's degree until he was 21. Shortly thereafter, on
1457 November 11, he was appointed as a member of the facul-
ty.

A few years later, he left Vienna for Rome, where he
arrived on 1461 November 20 and where he remained until 1463
July 5. During this period, he completed an epitome of
Ptolemy's Syntaxis that had been started by his colleague
Purbach, and which was based upon a medieval Latin translation of Ptolemy's work. Later in this period, Regiomontanus studied Greek in order to translate Ptolemy's work on
geography into Latin, but he never completed this project.
He also discovered an incomplete Greek manuscript of the
work of Diophantus, which may have been the beginning of our
recovery of this great mathematician's work.

Regiomontanus lived in Padua from the summer of 1463
until the summer of 1464. While there, he lectured and
wrote on the work of the early Islamic astronomers, and continued his research in trigonometry. He discovered the law
that connects the cosines of the sides of a right spherical
triangle, as well as what we call the law of sines for a
plane triangle. He subsequently wrote an important work on
trigonometry, both plane and spherical, and he prepared an
extensive table of sines which, perhaps for the first time,
had a decimal rather than a sexagesimal base. In fact, he
prepared sine tables with a base of 100 000 and, later, with
a base of 10 000 000. All of this work on trigonometry
remained unprinted until long after his death, but it apparently became well known during his life in manuscript
form.

At some time, perhaps in 1465, Regiomontanus moved to
Hungary, where he stayed until 1471. Here he finished his
sine tables, and began (and perhaps finished) his ephemerides of the solar system. Then, in 1471, he moved to
Nürnberg where he lived the rest of his life. Here he acquired a printing press and became, in Rosen's words,
". . the first publisher of astronomical and mathematical
literature, and he sought to advance the work of scientists

by providing them with texts free of scribal and typographical errors, unlike the publications then in circulation."

In particular, he published the first ephemeris table to be printed. This table gave the positions of the heavenly bodies for every day from 1475 to 1506, and Columbus is numbered among its users.

On 1471 June 2, Regiomontanus began a systematic set of astronomical observations in Nürnberg, which he continued until 1475 July 28, when he measured the meridian zenith distance of the sun. This observation provides our last certain knowledge of him. According to a contemporaneous record (see Rosen [1975]), the pope invited him to come to Rome and undertake a reform of the Julian calendar. If this is so, nothing came of it, because the calendar was not reformed until 1582, more than a century later. According to a sensational story that was first published in 1482, he was poisoned by the sons of a rival astronomer, and some sources give the exact date 1476 July 6 for his death.

There does not seem to be any factual basis for this account, and we do not know the date of his death. Rosen says that "in all probabiliity" he died in an outbreak of the plague in Rome that began in January of 1476. If so, he died when he was 39 or 40.

Bernard Walther (1430 - 1504) was a wealthy merchant who took up residence in Nürnberg in 1467, four years before Regiomontanus moved there. The most extensive account of him that I have seen is by Beaver [1970]. He was an amateur of astronomy in the original sense of the word. When Regiomontanus moved to Nürnberg, Walther supplied him with funds needed to acquire astronomical instruments, as well as with the printing press which Regiomontanus used to initiate the publication of accurate scientific texts.† Walther also did something that is considerably more important as far as the purposes of this work are concerned.

We have already noted that Regiomontanus made his last known observation on 1475 July 28. Five days later, Walther took up the series of observations that Regiomontanus had planned, and he continued them on a fairly regular basis for almost thirty years, until 1504 June 3. He died later in 1504.

Walther is known for two important innovations in observational astronomy. He is probably best known for being the first person to use a mechanical clock in making an astronomical observation: On the morning of 1484 January 16, he saw Mercury on the horizon. At that instant, he attached a weight to a clock which had 56 teeth in its hour wheel. When the center of the sun was on the horizon, the hour wheel had turned 1 full revolution and 35 teeth beyond,

---

†Tycho and Hevelius are other outstanding astronomers who had their own printing presses. See Hellman [1970] and North [1972].

so that the time interval was $1^h 37^m$. We do not know how the clock was regulated.†

Walther was also famous for his method of relating the longitude of a star to the sun and thence to the equinox. A method that goes back at least to the Greeks is described by Ptolemy [ca. 142, Chapter VII.2]. In this method, the observer measures the distance between the moon and the sun just before sunset. Then, just after sunset, he measures the distance between the moon and the star. He calculates the motion of the moon during the interval and thus he can find the angle between the star and the sun.

Walther realized that it is better to use Venus instead of the moon. There are two reasons why Venus is better: (1) It moves more slowly, so one can calculate its intervening motion more accurately; and (2) since Venus is essentially a point, we can locate it more accurately than we can the center of the moon.

I have used two printed forms of the combined observations made by Regiomontanus and Walther. The first form [Curtz, 1666] is often cited in the literature as Historia Coelestis. Sometimes this form is cited without giving any name for the editor. When a name is given, it may be Curtz, Barethis, Barrettus, Tycho, or Brahe. The correct name of the editor is Albert Curtz, who sometimes used the pseudonyms Lucius Barrettus and Lucius Barethio. Further, whether by design or inadvertence, he made it appear that the Historia Coelestis was written by Tycho Brahe, to which he (Curtz) contributed only a preface. The reader who wants to locate this work in a library may have to consult five different names before finding it in the catalogue.

The other printed form I have used is by Schmeidler [1972]. Unfortunately, on the title page of this edition, it is stated that this is the collected work of Regiomontanus, so that this edition is often cited under the name of the latter. Actually, this edition contains material by three people other than Regiomontanus, so I use Schmeidler's name in citing it. The reader who wants to locate this edition may have to refer both to "Regiomontanus" and "Schmeidler" in a library catalogue.

Schmeidler's edition is obviously composed of facsimiles of earlier printings. At least as far as the observations are concerned, it seems to be a copy of a work published by Johann Schöner in Nürnberg in 1544. This work is called Scripta Clarissimi Mathematici M. Ioannis Regiomontani, although it contains much material by others than

---

†The most accurate movement available at the time may have been the foliot movement that I have already mentioned; it had a torsion pendulum and escapement. It seems odd to me that the hour wheel should have had 56 teeth.

Regiomontaus.† I will have more to say about Schöner in the next section.

Schmeidler's edition is considerably more accurate than Curtz's, at least as far as the observations are concerned; I have not paid much attention to other parts of the works. Aside from this, there is a major difference in the arrangement of the observations. Curtz lists together all the observations made in a particular year. Within that year, he first lists all the observations of the sun in chronological order, followed by all other observations in chronological order, irrespective of the body observed. Schmeidler first brings together all the observations of the sun made by either Regiomontanus and Walther, in chronological order. This set of solar observations is followed by all of both men's observations of all the other bodies, in chronological order, again irrespective of the body observed.‡

With this information, I think the reader should have little trouble in locating any specific observation in either edition. Thus, in discussing a particular observation, I locate it by giving only the date and the body observed.

2. Walther's Observations That Were Used by Copernicus

Before taking up the observations made by Regiomontanus and Walther, I want to revert to a point that concerns Copernicus. As I mentioned in Section II.7, Copernicus uses three observations of Mercury made in his own time. He says that an observation made on 1491 September 9 was made by Walther in Nürnberg, and that observations on 1504 January 9 and 1504 March 18 were made by Johann Schöner, also in Nürnberg. ╪ There are two peculiarities about Copernicus's records of these observations.

First, as Beaver [1970] points out, all three of the observations of Mercury just mentioned were almost surely made by Walther. Walther records observations of Mercury on all three dates, and it seems unlikely that a different astronomer, also working in Nürnberg, made independent ob-

---

†I thank Professor Owen Gingerich, of the Harvard-Smithsonian Center for Astrophysics (private communication) for calling my attention to the publication by Schöner.

‡In both editions, the column of Walther's solar observations is preceded by a note saying that these are observations made per Bernardum Waltherum Nurenbergae discipulũ M. Ioannis de Monteregio.

╪Schöner (1477-1547) was a German astronomer and geographer. He is probably best known as being the first person to make a globe that used the name America. He came into possession of some of the papers of Regiomontanus and Walther after the death of the latter, and he published a work that was mentioned in the preceding section. He was apparently acquainted with Copernicus [Beaver, 1970].

servations of Mercury on two of the three dates. Thus, we should ask why Copernicus made this error in attribution.

TABLE III.1

WALTHER'S OBSERVATIONS OF MERCURY THAT
WERE USED BY COPERNICUS

Date			Walther				Copernicus			
			Longitude		Latitude		Longitude		Latitude	
			°	′	°	′	°	′	°	′
1491 Sep	9		163	23	1	50	163	30	1	50
1504 Jan	9		273	15	0	45	273	20	0	45
1504 Mar	18		26	30	3	00	26	6	3	00

Beaver points out that Copernicus may easily have been in Nürnberg in 1496, 1501, and 1503, although it is not certain that he was ever there. Still, if he were there, it is likely that he would have talked to Walther, who was already well known. If so, he could have obtained the observation dated 1491 September 9 at first hand. However, he apparently could not have been in Nürnberg in or after 1504, the year of the other two observations. It is plausible that he learned of these observations only from Schöner, and that he mistakenly assumed that Schöner was the observer. However, this is highly speculative and should not be relied upon.

The other peculiarity of these observations is shown by Table III.1. The first column in the table gives the date of an observation. Then come the longitude and latitude of Mercury that Walther actually measured, followed by the longitude and latitude that Copernicus says were measured. I have changed the fractions of degrees from the forms that Walther and Copernicus used to minutes of arc.

The latitude that Copernicus gives agrees with the latitude that Walther observed in every instance, but the longitude does not agree in a single instance. Kremer [1981] has studied this problem in considerable detail. He traces the values of the longitude stated by Copernicus through three different stages of development of his work and finds different values stated in each stage. He concludes that Copernicus "deliberately altered" the observed values to make them agree with his models, although others of his contemporaries preserved the observed values and tried unsuccessfully to make their models agree with the observations.

3. The Solar Observations Made by Regiomontanus and Walther

Even though it turns out that the solar observations made by these two astronomers are not directly useful for the purposes of this work, it is useful to study them briefly for the light they throw on other observations. I have studied the solar observations made by Regiomontanus and Walther in considerable detail in other work [Newton, 1982].

Regiomontanus made 29 measurements of the meridian zenith distance of the sun between 1472 and 1475, and Walther made 746 more between then and 1504. All these observations were made in Nürnberg. Walther's last 81 observations were made with an instrument that he calls a gnomon,† and all the others, by either man, were made with the type of instrument called a parallactic ruler.

Regiomontanus's 29 observations form a coherent series, made in one place with one instrument that was apparently not modified during the course of the observations. He made his first solar observation in Nürnberg on 1472 February 20 and his last one on 1475 July 28, and we noted in Section III.1 that the record of this observation provides our last certain knowledge of him.

Walther took up the series of observations five days later, on 1475 August 2. It is not safe to assume that Walther took over the instrument with no modification; for instance, he might have taken it to a different house. Thus we must consider that a new series of observations begins on 1475 August 2. After this date, the observations continue on a fairly steady basis until 1479 September 19, but then there is a gap until 1487 June 10. We cannot assume that Walther made no change in his parallactic ruler over this long period, so we must consider that the observations from 1475 August 2 to 1479 September 19 form a second group.

A third group runs from 1487 June 10 to 1496 September 15. The next observation, made on 1496 September 17, has a note that it was made with a new instrument, presumably a new parallactic ruler. Thus, a fourth group of observations starts on this date, and continues until 1501 December 10.

There are no observations between 1501 December 10 and 1503 March 15, and the record of this date says that the observation was made in a new house. Hence, we must start a fifth group of observations on this date; the fifth group continues through 1503 August 10. On the next day, we find a note that the instrument had been rectified, and from then through 1503 September 8 we have a sixth group forming a strange mixture of observations. Walther modified the parallactic ruler several times, and he mingled observations made with the ruler and observations made with his gnomon.

---

†I have shown that this instrument was not like the instrument that is generally called a gnomon. Instead, it was like a meridian quadrant arranged so that one read the chord of the angle rather than the tangent.

This sixth group clearly does not form a coherent set, and we cannot use it. I will ignore it in the rest of the discussion.

The seventh and last group begins on 1503 September 10 and runs to 1504 June 3. This date provides the last known observation by Walther. The seventh group was made entirely with the gnomon, according to Walther's statement.

For each of the six usable groups, I formed an estimate of $\Delta\underline{T}$, the difference between dynamical time and solar time. I also estimated the bias in setting the local vertical, a parameter ε that measures seasonal effects, and the standard deviation of the residuals. The parameter ε needs some explanation.

Suppose, for example, that the alignment of the instrument depends upon the temperature and, therefore, on the average, upon the season. Then, if the declination of the sun changes by 47°, say, between the winter and summer solstices, the declination (change in zenith distance) as measured by the instrument changes by a different amount. If $\Delta\delta$ denotes the change in declination, I let $(1 + \varepsilon)\Delta\delta$ denote the change measured by the instrument.† I found that ε is negative for all groups of observations, and that the values average about -5×10^{-4}. This means that the total swing in declination indicated by the instruments is less than the true swing by about 80″. Part of this value may come from an inaccurate modelling of the atmospheric refraction.

The standard deviation of the residuals is about 95″ for the group of observations made by Regiomontanus and about 55″ for all groups of observations made by Walther. The performance of his gnomon was essentially the same in all respects as the performance of his parallactic rulers. If the residuals were random, we could get the position of the sun to an accuracy of about $55''/\sqrt{746} = 2''$, and the observations would give us a valuable estimate of the earth's spin acceleration.

Unfortunately, inspection of the residuals shows that they are not random. Instead, the residuals remain correlated over substantial periods of time, even though we have presumably removed the seasonal effect to first order by means of ε. When we estimate the value of $\Delta\underline{T}$, and when we allow for the correlations in attaching a standard deviation to it, we get

$$\Delta\underline{T} = 727 \pm 1310 \quad \text{seconds.} \qquad \text{(III.1)}$$

The corresponding estimate of the earth's spin acceleration $\underline{y}$ is

---

†This process removes an effect with a period of a year that is in phase with the seasons, but it cannot remove an effect that is out of phase. Thus, I could not use it with Tycho's data. See Section II.6.

$$\underline{y} = -50.7 \pm 90.7 \qquad\qquad (III.2)$$

parts in 10^9 per century. These estimates apply at the mean
epoch of the observations, which is 1487 November 15. For
comparison, we expect $\Delta\underline{T}$ to be about 170 seconds.

I believe that the error estimate in Equation III.1
furnishes us for the first time with a realistic estimate of
the accuracy of ancient and medieval observations of the
sun. This estimate should apply reasonably well to many
collections of observations, and the error estimate for a
collection should be reasonably independent of its size,
provided that it contains a moderate number of observations
spread over a considerable time.

For observations made around the year 1000, we expect
the value of $\Delta\underline{T}$ to be about 1300 seconds. If this is so,
the error in finding $\Delta\underline{T}$ from measurements of the sun's declination is greater than the central value for any observations made since 1000. In order to get the earth's spin
acceleration, then, we must either use older observations or
use a different type of observation. In particular, it is
not likely that we can get a useful estimate from any solar
table newer than that of ibn Yunis [1008].

4. Observations of Lunar Eclipses Made by Regiomontanus
and Walther

The study described in the preceding section makes it
unlikely that we can form a useful estimate of an astronomical acceleration by using direct observations of position
made by Regiomontanus and Walther. Their only observations
that are likely to be useful are their observations of
eclipses. I will analyze their observations of lunar eclipses in this section and their observations of solar eclipses
in the next one. I start by giving a brief summary of their
observations of lunar eclipses.

1457 September 3. This was a total eclipse. At the
beginning of totality, the altitude of the next to the last
of the Pleiades was 22° east. At the end of totality, the
altitude was 36°. This observation was made at Melk Castle
in Austria.

1460 July 3. Regiomontanus identifies this as a partial lunar eclipse in the night that followed July 3; we
know from calculation that its center was well before midnight. He says that its beginning was at 7^h 16^m after noon,
its middle was at 8^h 13^m, and its end was at 9^h 10^m, according to tables prepared for Vienna. He observed that the
magnitude was 4 digits† and some more. At the end of the
eclipse, the altitude of the moon was 15° 18 '. His "precep-
tor" George helped make this observation in Vienna.

---

†A magnitude of 1 digit means that 1/12 of the diameter was
  observed. This definition applies to both solar and lunar
  eclipses.

1460 December 28. This was a total eclipse, which
Regiomontanus dates December 27. Actually, its center was
somewhat after 1:00 AM on December 28 in Vienna, where
Regiomontanus observed it. At the beginning of the eclipse,
the altitude of "Alramech" was 7° to the east. At the beginning of totality, the altitude of the star was 17° and at
the end of totality the altitude was 28°. At the beginning
of the eclipse the moon was in a great circle with α Geminorum and α Canis Minoris. At the end, the moon was in a
great circle with β Geminorum and α Canis Minoris. I will
not use these statements of lunar position.

1461 June 22. This was also a total eclipse. At the
beginning of totality, the altitudes of α Aquilae and the
moon were 26° and 6° 30', respectively. At the end of the
entire eclipse, the altitudes were 47° 30' and 17° 30'. The
place is Vienna.

1461 December 17. On this evening, the moon rose
eclipsed by 10 digits of its diameter, but Regiomontanus saw
"exactly" 8. At the end of the eclipse, the altitude of
"Alhaioth" was 38° 30' east and that of Aldebaran was 29°
east. The observation was made in Rome. Eight digits was
certainly not the maximum magnitude of the eclipse, which
was, in fact, total, and it is not clear exactly when the
magnitude had this value. Hence, I will use only the timing
of the end of the eclipse and ignore the statement about the
magnitude.

1462 June 12. Regiomontanus identifies this as a partial eclipse in the night that followed June 11†; its center
was at nearly three hours after midnight. He was in Viterbo. He could not see the beginning of the eclipse because
of clouds. At the middle of the eclipse, by "conjecture",‡
the altitude of α Aquilae was 51° west. Seven digits were
"estimated"‡ to be eclipsed. In this context, both "conjecture" and "estimate" could refer either to a measurement of
poor quality or to a calculation. I will not use this record; if the figures given did come from observation, it is
likely that the quality of the observation was impaired by
clouds.

1464 April 21. This was a total eclipse. Regiomontanus gives a number of calculated quantities for this
eclipse, but he measured only one of them. At the beginning, he says, the altitude of α Scorpii was 12° 45' east
and that of α Hydrae was 9° 40' west. The observation was
made in Padua.

---

†Incidentally, Regiomontanus uses the modern style of designating the days of the month, and he does not use either
the Roman or Italian calendars.

‡"per coniecturam".

‡"putabantur".

1471 June 3. Regiomontanus dates this as June 2. At the beginning of the eclipse, the altitude of α Scorpii was 14° 15' west, while the altitude of Delphin or Muscida Pegasi was 22° 30' east. Four digits were obscured. This was observed in Nürnberg.

1487 February 8. The middle of the eclipse came when the altitude of the nadir of the sun was 29°; this time was $3^h 47^m$. At the end of totality, the altitude of the nadir of the sun was 24°, and the time was $4^h 18^m$. The time at the end of the eclipse was $5^h 20^m$. Walther says that the times were determined by altitudes, but he does not say of what. These observations were made in Nürnberg.

TABLE III.2

MAGNITUDES OF LUNAR ECLIPSES MEASURED BY REGIOMONTANUS

Date	Measured magnitude, digits	Calculated magnitude, digits	
		$\dot{n}_M = -28$	$\dot{n}_M = -18$
1460 Jul 3	4+	3.46	3.36
1471 Jun 3	4	7.39	7.31

1504 March 1. Walther could see neither the beginning of this eclipse, nor the beginning of totality, nor the end of totality, because of clouds. At the end of the eclipse, the middle of the sky was at 10° of Scorpius. Since conditions do not seem to have been favorable, I will omit this record.

These records contain statements about the magnitudes of only four eclipses, and two of these are the unusable measurements of 1461 December 17 and 1462 June 12. In contrast, there are 14 measurements of time associated with lunar eclipses. Much of the reason for this difference is, of course, that there is no way to measure the magnitude of a total eclipse, only of a partial eclipse. The analysis of the magnitudes is simpler than that of the times, and I take it up first.

Table III.2 gives the pertinent information about the two measurements of magnitude that are safe to consider. The first column in the table gives the date of the eclipse and the second column gives the measured magnitude in digits; 12 digits constitutes totality. The final two columns give calculated values of the magnitude; these need discussion.

It is clear that the magnitude of a lunar eclipse depends upon the orbital positions of the sun and moon, but that it does not depend upon the orientation of the earth. In other words, the magnitude is a function of $\dot{\underline{n}}_M$ but not of $\underline{y}$ or of ΔT. I have calculated the magnitude using two values of $\ddot{\underline{n}}_M$, namely -28 and -18 seconds of arc per century per century.

The table shows us immediately that the measurements of magnitude will make no useful contribution to the estimate of $\dot{\underline{n}}_M$: The dependence of the magnitude upon the acceleration is small and the errors in the measured values are large. The errors need a little further discussion.

Earlier [Newton, 1970, p. 220], I found that the standard deviation of a naked-eye measurement of the magnitude of a lunar eclipse is about 0.07; this is about 0.84 digits. I found that this is so for all sets of such measurements I studied except for the Greek observations reported by Ptolemy, which showed a puzzlingly low scatter. At that time, I had not discovered that these data were fabricated. Now that we know this, the low scatter is no longer a puzzle.

The error shown in Table III.2 for the eclipse of 1460 June 3 is in line with the expected standard deviation of about 0.84 digits. The error for the other eclipse, however, is over 3 digits, about 4 standard deviations. It is likely that the recorded value for this eclipse is a scribal error. If so, the error was probably made by Regiomontanus himself, since the value is the same in both texts that I have used.

Now we can turn to the measurements of time associated with lunar eclipses. If the reader will go back to the records summarized above, he will see that the times were measured by measuring the altitude of a star, of the moon, or in one case of the nadir of the sun; it is not clear that the latter is actually a measurement. The stars used pose some problems.

1457 September 3. In this observation, Regiomontanus uses the "next to the last" of the Pleiades. There are six Pleiades in mythology and I believe that six is the number a person with reasonably good vision can see. Thus it is puzzling that there are only four Pleiades in Hipparchus's star catalogue, which Ptolemy [ca. 142] has transmitted to us, and there are also only four in Tycho's catalogue. I have studied the latter catalogue carefully but I have not published the study. In the context of the observation, "last" means having the greatest right ascension, so that Regiomontanus uses the one of the Pleiades with the next to the largest value of right ascension.

There is considerable confusion about which stars constitute the four Pleiades in the old catalogues. Luckily, in my opinion, there is little question about the star meant here. People with keen vision see seven stars in the

Pleiades, which are the stars 27, 28, η, 23, 20, 19, and 17
Tauri. The stars 27 and 28 are quite close together, and I
believe that people who see only six Pleiades see these two
as a single star. If so, this "star" is the one with the
largest right ascension. If we count 27 and 28 Tauri as a
single star, the one of the Pleiades with the next largest
right ascension is η Tauri, and it is also the brightest
star in the Pleiades. Thus, I think there is little doubt
that the "next to the last" of the Pleiades is η Tauri.

1460 December 28. In this record, Regiomontanus uses
the star "Alramech". This sounds like an Arabic name for a
star, but I have not been able to find this name in any list
of named stars. In an attempt to identify the star, I calculated the position of Regiomontanus's zenith at the calculated time of the beginning of the eclipse and marked this
point on a globe. Then I marked the circle that was 83°
from the zenith (corresponding to an altitude of 7°) and
noted the reasonably bright stars that lie close to this
circle. The only likely possibilities I could find are α
Bootis and η Bootis, but neither of these fits the data well
enough to give us confidence in an identification. I will
not use the observations of this eclipse.

1461 June 22. In the record of this eclipse, Regiomontanus gives the altitudes of α Aquilae and of the moon at
the beginning of totality and at the end of the eclipse.
The latter altitude of the moon is 17° 30'.

At the time of the eclipse, the declination of the moon
was about -24° and the observation was made at a latitude of
about 48°. Thus, the maximum altitude of the moon the night
of the eclipse was only about 18°. We cannot determine the
time with useful precision from an altitude so close to the
maximum, and I will not use the altitude of the moon at the
end of the eclipse.

1461 December 17. In this record, Regiomontanus uses
stars named "Alhaioth" and "Aldebaran". There is no trouble
about Aldebaran; it is α Tauri. In the list of named stars
in my dictionary, there is a star named Alioth; it is ε
Ursae Majoris. However, this star is nowhere near the required place, and I have found no bright star that is. From
necessity, then, I use only α Tauri in analyzing this observation.

1471 June 3. Here, Regiomontanus uses a star that he
designates "Delphin or Muscida Pegasi". I cannot find
either name attached to a star. Turning first to Delphin,
there is a small constellation close to Pegasus named Delphinus, and the unaccompanied name of a constellation was
often used to designate the brightest star in the constellation. This suggests that the star α Delphini might once
have been associated with Pegasus and that it might be the
star in question. However, neither α Delphini nor any other
star in Delphinus could have had the altitude given in the
record. I believe that the star is either β or ε Pegasi.
These stars are separated by nearly 30° on a great circle,
but their elevation angles at the times of the observation
are almost the same. Thus, in analyzing this observation, I

will use the average of the times determined by using β and ε Pegasi. The times differ by about a minute, so the average is in error by only about 30 seconds.

TABLE III.3

A COMPARISON OF INFERRED AND STATED TIMES
FOR THE ECLIPSE OF 1487 FEBRUARY 8

Phase of eclipse	Hour inferred from altitude of sun's nadir	Stated hour
Middle	3.95	3.78
End of totality	4.48	4.30

1487 February 8. At two phases of the eclipse, namely the middle of the eclipse and the end of totality, Walther gives both the time and the altitude of the nadir of the sun. I cannot reconcile the altitudes and the stated times, whether I assume that the stated times are mean or apparent local time. The situation is shown in Table III.3. In the table, the first column gives the phase involved, and the second column gives the local apparent time calculated from the altitude of the sun's nadir. The third column gives the time that Walther states. The local mean time is almost exactly 0.25 hours later than the apparent time given in the second column.

The time inferred from the altitude is greater than the stated time by almost the same amount for both phases, and the discrepancy is greater if the time is taken to be mean than if it is taken to be apparent.

The nadir of the sun is not an observable point, and its position can be found only from calculation. I do not understand the reason why Walther gives the altitude of the nadir, but I believe that he calculated it for some reason and made an error in doing so. Perhaps he had a poor table of the sun.† I will assume that the stated hours are correct, since Walther says explicitly that he found them from altitudes.

The middle of an eclipse cannot be observed directly with precision. It is safe to assume that Walther knew this and that he found the time of the middle by some kind of averaging, perhaps of the times of the beginning and end of totality.

---

†It would be interesting to do the calculations using the ephemeris that Regiomontanus printed; this is probably the ephemeris that Walther used.

When I use the altitude of any object to find the time, I first correct the measured altitude for refraction. The lowest apparent altitude that concerns us is 6° 30', the apparent altitude of the moon in the eclipse of 1461 June 22 as observed in Vienna. Refraction changes this altitude by about 0°.13, and this changes the inferred time by less than a minute. Hence, we do not need to calculate the refraction with great care.

To find the refraction, I use the table of Allen [1962].† This table gives the refraction under the standard conditions of 10°C and a pressure of 1 atmosphere. Since the refraction is not needed with great accuracy, it is sufficient to use the standard table, without attempting to correct it for the actual values of temperature and pressure at the time of each measurement.

Next, I turn to the ephemerides of the various objects that were used to time the observations. I have ephemeris programs that first give the right ascension and declination, with respect to the true equinox and equator of date, of the sun, moon, planets, and about 1000 stars, for any specified value of Greenwich mean time (solar time); the precision of these programs is a fraction of a second of arc. The programs include the effects of aberration and parallax. The positions found for the sun, moon, and planets depend upon the secular accelerations assumed. However, for the time of the observations made by Regiomontanus and Walther, the effect of the accelerations upon the position of an object is less than a minute of arc, and the effect upon the inferred time is a few seconds.

Hence, when we use the position of some object only to furnish the time, it is sufficient to use provisional values of the accelerations, and it is not necessary to consider how the inferred time depends upon the accelerations. Thus, in calculating the times from the measured altitudes, I use provisional accelerations found from earlier work; these are the accelerations presented in Section I.5.

After a program finds the right ascension and declination of the object in question, it then uses the latitude and longitude of the observer to find the azimuth and altitude of the object as seen by the observer. Thus, I enter the ephemeris program of the object in question with two times that span the time of the observation. After finding the right ascension, declination, azimuth, and altitude for the two times, I interpolate to find the time at which the calculated altitude equals the measured one. By the measured altitude, I mean the altitude after correcting for refraction.

---

†In Section II.6, I calculated the refraction using Equation II.5, but there I was able to ignore altitudes less than 20°. Here, we must use altitudes down to 6° 30', so we must use the tabulated refraction instead of the equation.

The latitudes and longitudes of the places involved in the observations are listed in Table A.III.1 in Appendix III. I believe that the use of this table is obvious.

TABLE III.4

SOME TIMES INFERRED FROM MEASURED ALTITUDES

Date	Object observed	Measured altitude °	Mean solar time h
1457 Sep 3	η Tau	22	21.356
	η Tau	36	22.760
1460 Jul 3	Moon	15.3	21.131
1461 Jun 22	α Aql	26	19.730
	Moon	6.5	19.796
	α Aql	47.5	22.681
1461 Dec 17	α Tau	29	16.614
1464 Apr 21	α Sco	12.75	22.330
	α Hya	9.67	22.253
1471 Jun 3	α Sco	14.25	-1.067
	β Peg	22.5	-1.038
	ε Peg	22.5	-1.018

The times found this way are given in Table III.4. In this table, the first column gives the date of the eclipse and the second column gives the celestial object used to furnish the time. The third column gives the altitude of the object recorded by Regiomontanus or Walther, and the fourth column gives the hour inferred from the altitude.

In all cases, I have used the date of an eclipse that is assigned in Oppolzer's Canon [Oppolzer, 1887]. This has required me to use negative values of the hour for the eclipse of 1471 June 3.

If the sun had been used to furnish the hour, the hour found would be mean solar time (at Greenwich), as the heading of the fourth column says. When some other object is used, the inferred time can differ from solar time by the error made in calculating the position of the object with respect to the sun, divided by the diurnal rate of rotation of the earth. At the period of the observations in Table III.4, the error in relative position is of the order of a minute of arc or less, so the difference between solar time and the inferred time is of the order of a few seconds or

less. Thus, I consider all the times to be (mean) solar
time.

TABLE III.5

OBSERVATIONS OF LUNAR ECLIPSES BY REGIOMONTANUS AND WALTHER

Date	Phase	Measured time $\underline{h}$	Calculated time[a] $\underline{h}$	$1000 \times \partial\underline{h}/\partial\underline{y}$	$1000 \times \partial\underline{h}/\partial\underline{\dot{n}}_M$
1457 Sep 3	BT	21.356	21.469	4.8	−5.5
	ET	22.760	22.645	4.9	−6.8
1460 Jul 3	E	21.131	21.057	4.7	−9.5
1461 Jun 22	BT	19.730[b]	19.791	4.8	−7.5
	BT	19.796[c]	19.791	4.8	−7.5
	E	22.681	22.619	4.7	−7.7
1461 Dec 17	E	16.614	16.609	4.6	−6.2
1464 Apr 21	B	22.230[d]	22.296	4.6	−5.7
	B	22.253[e]	22.296	4.6	−5.7
1471 Jun 3	B	−1.067[d]	−1.066	4.3	−5.9
	B	−1.028[f]	−1.066	4.3	−5.9
1487 Feb 8	M	3.294	3.312	4.0	−5.6
	ET	3.811	3.880	4.0	−5.0
	E	4.844	5.004	4.0	−5.5

[a]Using $\dot{n}_M$ = −28 and $\underline{y}$ = −20.

[b]Using α Aquilae.

[c]Using the moon.

[d]Using α Scorpii.

[e]Using α Hydrae.

[f]Using the average of β and ε Pegasi.

Next, we want to calculate the times in question from
modern theory for some standard set of the accelerations, as
well as the derivatives of the times with respect to the
accelerations. These calculated quantities are shown in
Table III.5. In the table, the first column is again the
date of the eclipse. The second column shows the phase of
the eclipse whose time was measured. In this column, I use
the following abbreviations:

B: beginning of the eclipse,

BT: beginning of totality,

M: middle of the eclipse,

ET: end of totality,

E: end of the eclipse.

The third column gives the measured times of the stated
phases; these are taken from Table III.4, with a few exceptions: The second time given for 1471 June 3 is the average
of the times found using β and ε Pegasi. The times given
for the eclipse of 1487 February 8 are those stated by
Walther, converted from local apparent time to Greenwich
mean time.

The results of the calculations are given in the remaining columns. The fourth column gives the time of the
stated phase calculated using $\underline{n}_M = -28$ and $\underline{y} = -20$. The
fifth column gives 1000 times the derivative of the hour h
with respect to $\underline{y}$, and the last column gives 1000 times the
derivative of $\underline{h}$ with respect to $\underline{\dot{n}}_M$

In three cases in Table III.5, the time of a particular
phase is deduced from the altitudes of two different celestial objects. Such observations are not entirely independent, and we must decide how to treat them before we use
the table. We have two extreme cases. In one extreme, the
observer can judge precisely when a particular phase occurs
but he makes a large error in measuring the altitudes from
which the time is inferred. In the other extreme, he can
measure the altitudes precisely but he makes a large error
in judging when the particular phase occurs. In the first
extreme, we should take the observations to be essentially
independent. In the second, we should treat the pair of
times as a single observation, using the average of the
tabulated times in the analysis.

When we have two measurements of the same phase, the
difference between the two times ranges up to 0.066 hours,
or about 4 minutes. We will see in a moment that this is
comparable to the accuracy with which the observers could
judge when a phase occurred, and this means that we have a
complicated situation lying between the extremes. Luckily,
the final inference does not depend much upon how we weight
the sets of dual observations, so I make the simplifying
choice of taking them to be independent.

I will not use the derivatives with respect to $\underline{\dot{n}}_M$ in
the analysis, for a reason that was explained in Section
I.7, and I will estimate only the acceleration $\underline{y}$. The
least-squares solution for $\underline{y}$ is

$$\underline{y} = -23.2 \pm 4.3. \hspace{3cm} (III.3)$$

The standard deviation listed is the formal standard deviation that results from the scatter in the data. That is, it

is the precision of the result rather than its accuracy. The reader can easily get the standard deviation from Table III.5 with the additional information that the residual for a particular observation is almost exactly the same as the difference between the third and fourth columns in the table.

However, there is almost surely some systematic error in the measured times in Table III.4, and for observations as recent as these a systematic error can be important. The size and sign of the systematic error depends upon the work habits of the observer. For example, he might have watched the moon carefully for the beginning of totality. Then, when he thought he saw totality, he turned to his quadrant or whatever he used and measured one or more altitudes. In this case, his times would be systematically late by the time it took him to measure an altitude.

On the other hand, he might have made systematic records of the altitude of one or more objects while waiting for totality to begin. Then, when he thought he saw totality, he took the time to be that of the latest altitude he had recorded. In this case, his times would be systematically early by about half the interval between measurements.

I think a minute is a reasonable estimate of the order of the systematic error in the observations.† Changing all the times by a minute makes a change of about 3.6 in the value of $\underline{y}$. Arbitrarily, I add this to the standard deviation in Equation III.3 and take

$$\underline{y} = -23.2 \pm 7.9 \qquad\qquad \text{(III.4)}$$

as the best estimate we can make from the lunar eclipses observed by Regiomontanus and Walther. The epoch for Equation III.4 can be taken as 1472.

5. Observations of Solar Eclipses Made by Regiomontanus and Walther

Now we turn to the observations of solar eclipses that Regiomontanus and Walther made, beginning with a brief summary of the records.

1462 November 21. Regiomontanus could not see the beginning of the eclipse. When two digits of the diameter were eclipsed from the southern side, the solar altitude was 26 $\frac{1}{2}$ degrees. At the end, which he carefully noted, the altitude of the sun was 24° 36', and its azimuth was 16° 15' from south to west. The eclipse was observed in Viterbo.

---

†However, astronomers five centuries ago had no tables that were accurate enough for a minute to matter. Thus, they may not have tried to measure times as precisely as a minute.

This eclipse was near the middle of the day in Viterbo. If Regiomontanus did not see the beginning, it might have been due to clouds or, perhaps, to some obligation that prevented him from observing.

The first observation is impossible. At the center of the eclipse, the declination of the sun was about $-21°.73$, and the latitude of Viterbo is $42°.40$. Thus, the meridian altitude of the sun was less than 26°. We cannot use this observation.

At the end of the eclipse, the altitude of the sun was still only about 1° less than its maximum value, and the altitude determines the time with low precision. The only useful datum from this record is the azimuth of the sun at the end of the eclipse. It is disconcerting that an astronomer as gifted as Regiomontanus would attempt to use an altitude to give the time under these circumstances.

1473 April 27. This record does not occur in its expected place. Instead, we find it inserted among the meridian elevations of the sun. The elevation of the sun was 25° at the end of the eclipse; the meridian elevation is not given. This notice is found on page 627 of Schmeidler's edition [Schmeidler, 1972]. Luckily, the meridian elevation was about 57° on this day, and the altitude gives the time with reasonable precision. This observation and all the remaining ones were made at Nürnberg.

1478 July 29. At the beginning of the eclipse, the altitude of the sun was $54 \frac{1}{2}$ degrees and at the end the altitude was $41 \frac{1}{2}$ degrees. Again, the meridian altitude was about 57°, so the first observation is not useful. Only the second will be used. The preceding observations have been by Regiomontanus. This and the remaining ones are by Walther.

1485 March 16. Walther says that the time was $3^h 26^m$ at the beginning of this eclipse, and that it was $5^h 28^m$ at the end. The magnitude was about 11 digits, although by calculation the eclipse should have been total.

Here Walther gives the times with a precision of a minute without giving the data from which the times were found. He did the same thing for the lunar eclipse of 1487 February 8 in the preceding section, and those times seemed to be as accurate as any that we calculate from measured altitudes. Tentatively, the main risk in using times stated by Walther is that they might have been calculated rather than observed. Here, Walther observed the magnitude and found that it disagreed with the calculated one. This gives some grounds for feeling that the times may also be observed.

I will adopt a cautious attitude toward these times and those in the remaining records. If the tables were so much in error that they indicated a total eclipse, the calculated times will probably be off enough that we can find this out by analysis. Thus, I will study the recorded times and adopt them if the errors seem to be reasonable. I will also see if the reading of the magnitude has enough precision to be useful.

 1491 May 8. All Walther says of this eclipse is that the beginning came when the middle of the sky was 19° of Gemini and that the end came when the middle of the sky was 26° of Cancer. I presume he means that these points on the ecliptic were in the meridian at the beginning and end of the eclipse. These points cannot be observed directly, certainly not in the daytime, and referring to them is an oblique way of stating times that must have been found some other way. The time at the end should be reasonably accurate if Walther found it from either the altitude or the azimuth of the sun. The time at the beginning will be useful only if he found it from the azimuth.

TABLE III.6

MEASUREMENTS OF TIMES CONNECTED WITH SOLAR ECLIPSES

Date	Phase of eclipse[a]	Quantity observed	Local apparent time h	Equation of time[b] h	Local mean time h
1462 Nov 21	E	Azimuth S	13.063	+0.173	12.890
1473 Apr 27	E	Altitude S	7.285	+0.067	7.218
1478 Jul 29	E	Altitude S	14.950	−0.080	15.030
1485 Mar 16	B	Clock	15.433	−0.102	15.535
	E	Clock	17.467	−0.102	17.568
1491 May 8	B	79° of ecliptic[c]			13.548
	E	116° of ecliptic[c]			16.205
1493 Oct 10	E	271° of ecliptic[c]			16.192
	E	Clock	16.400	+0.247	16.153
1497 Jul 29	E	Clock	15.400	−0.080	15.480

[a]B = beginning of eclipse; E = end of eclipse.
[b]In the sense of apparent time minus mean time.
[c]The point of the ecliptic that was in the meridian.

1493 October 10. Walther says that conditions were
poor at the beginning of the eclipse. At the end, the middle of the sky was 1° of Capricorn, namely, $4^h 24^m$ (after
noon). Here, as with the lunar eclipse of 1487 February 8,
we have a chance to test the consistency of two ways that
Walther used to state the time.

1497 July 29. Walther could not see the beginning of
this eclipse either. The end was at $3^h 24^m$ in the afternoon.

Before we turn to the times taken from these records,
let us consider the single observation of magnitude, on 1485
March 16. Walther said that the magnitude was about 11
digits, although the eclipse was total by calculation.
Calculation of the magnitude using $\dot{n}_M = -28$ and $\bar{y} = -20$
yields 11.71 digits, and changing $\bar{y}$ by 10 changes this
figure by only 0.03. It is clear that this estimate of the
magnitude does not give a useful estimate of $\bar{y}$, although
Walther is correct in saying that the eclipse was not total.

The times from the records that will be considered are
summarized in Table III.6. The first column gives the date
of the eclipse and the second designates the phase whose
time was measured; this is the end in all but two cases.

TABLE III.7

OBSERVATIONS OF SOLAR ECLIPSES BY REGIOMONTANUS AND WALTHER

Date	Phase	Measured time h	Calculated time[a] h	1000 × $\partial h/\partial y$	1000 × $\partial h/\partial \dot{n}_M$
1462 Nov 21	E	12.890	12.917	6.1	-6.4
1473 Apr 27	E	7.218	7.260	5.2	-8.2
1478 Jul 29	E	15.030	15.076	4.9	-7.0
1485 Mar 16	B	15.535	15.555	4.8	-6.3
	E	17.568	17.589	4.3	-5.7
1491 May 8	B	13.548	13.427	5.7	-9.6
	E	16.205	16.273	4.7	-8.3
1493 Oct 10	E	16.192[b]	16.232	4.5	-7.8
	E	16.153[c]	16.232	4.5	-7.8
1497 Jul 29	E	15.480	15.729		

[a]Using $\dot{n}_M = -28$ and $\bar{y} = -20$.
[b]Using the point on the ecliptic that was in the meridian.
[c]Stated clock time.

The third column tells us how the time was determined. The azimuth of the sun was used in one case and its altitude in two. When I use the altitude, I correct it for refraction before inferring the time. The time found from the sun is local apparent time, whose value is given in the fourth column. To compare with calculated values, we change this to local mean time, using the equation of time tabulated in the fifth column. The equation of time is given in the sense of apparent time minus mean time. The final column gives the value of the local mean time. In contrast to what I did with the lunar eclipses, I leave this as local time instead of converting to Greenwich time.

In four cases, Walther gives the time without stating how he found it. I have put "clock" in the third column of Table III.6 in these cases, although it is not necessary that Walther actually found the times by reading a clock. I imagine that he found them from the position of the sun, although it is certainly possible that he set a clock by the sun at some earlier time and then used the clock for the duration of the eclipse. In any case, it seems safe to assume that these times are local apparent time. It is also safe to assume that they are measured and not calculated.

In the remaining three cases, Walther states the point on the ecliptic that was in his meridian. If he calculated this point carefully, the time given must be local mean time rather than apparent time, and the time derived from it must be local mean time rather than apparent time, and I have assumed that it is mean time in preparing Table III.6. To find the time in these cases, I first calculated the right ascension of the stated point on the ecliptic and the right ascension of the fictitious mean sun. The difference between these is the hour angle of the fictitious mean sun, which, of course, gives us the local mean time immediately.

The analysis of the times is summarized in Table III.7. The first column gives the date, the second column gives the phase that was observed, and the third column gives the local mean time of the observation; these columns are taken from Table III.6. The fourth column gives the time of the observed phase as calculated using the DE102 ephemeris, modified to make $\underline{n}_M = -28$ and $\underline{y} = -20$. The last two columns give the partial derivatives of the hour $\underline{h}$ with respect to $\underline{y}$ and $\underline{n}_M$. I do not use the latter derivative, but I include it for the sake of any reader who may want to use it.

We see immediately that the measured time for the eclipse of 1497 July 29 is inconsistent with the others. Since the measured time reads the same in both texts, Walther must have made an error either in finding the time or in recording it. (Perhaps it is a calculated rather than an observed time.) I do not use this observation, and I indicate this lack of use by omitting the partial derivatives for this observation.

When we infer the value of $\underline{y}$ in the usual way, we get

$$\underline{y} = -24.2 \pm 4.2. \tag{III.5}$$

This is remarkably close to the value found from the lunar eclipses in Equation III.3. As I did before, I increase the standard deviation in Equation III.5 by 3.6 to allow for the possibility of a systematic error of a minute in measuring the times. Thus, the final estimate I make from the times of solar eclipses is

$$\underline{y} = -24.2 \pm 7.8. \hspace{3cm} \text{(III.6)}$$

The epoch for Equation III.6 will be taken as 1480.

EUROPEAN AND CHINESE OBSERVATIONS BETWEEN 700 AND 1400

1. European Astronomy in the Middle Ages

It is often written that interest in Europe† in the sciences, as well as interest in ancient learning in general, arose as a consequence of the Crusades. The Crusades, it is said, were the agency that first brought Europeans into contact with ancient learning, which had been preserved by the Islamic peoples.

So far as astronomy is concerned, at least, I suspect that this idea is not correct. It is indeed generally true that Islamic astronomy grew out of Greek astronomy and that European astronomy grew out of Islamic astronomy. However, a serious European interest in astronomy began before the Crusades. Further, it is not likely that most of the transfer of learning from Islam to Europe took place in the war zone around the eastern Mediterranean. It is likely that most of the transfer took place in Spain and Sicily, where Islamic conquerors and their European subjects lived in relative peace for substantial periods of time.

As a notable example of European interest in science before the Crusades, Gerbert of Auvergne (later Sylvester II, pope 999-1003) was writing extensively on mathematics and science by the year 1000. As another example, Walcher of Malvern used an astronomical instrument to observe the lunar eclipse of 1092 October 18.‡ He used this observation, and probably others that have not survived, as the basis for calculating tables of the motion of the moon [Haskins, 1924, pp. 114ff]. I will analyze his observation in Section IV.6 below.

The first European activity, other than observation, seems to have been mostly translation. Adelard of Bath (fl. 1130) and Gerard of Cremona (1114?-1187), for examples, prepared Latin translations of famous Arabic writers such as al-Khwarizmi,‡ and famous Greek writers such as Ptolemy, Euclid, Aristotle, and so on. The first translations of

---

†When I write "Europe" in this context, I probably mean "Roman Catholic Europe". So far as I know, the parts of Europe under the Greek Orthodox Church displayed little interest in the sciences. Likewise, so far as I know, such areas as Scandinavia showed no interest in science until after their conversion to the Catholic faith. I exclude those parts of Europe that were under Islamic domination, as belonging to a different culture. This seems to leave only Catholic areas.

‡The First Crusade began in 1096.

‡See Sections V.1 and V.6.

Greek writers were double translations – from Greek to Arabic to Latin – but later ones were made directly from the original language.

Later scholars such as Sacrobosco (fl. 1230), Albertus Magnus (1206?-1280), Thomas Aquinas (1225?-1274) and Roger Bacon (1214?-1294), mastered the materials that translation had brought to them, and prepared commentaries and textbooks based upon them. With the latter, Pannekoek [1961, p. 176] writes that European science at last came up to the level of classical antiquity. Henceforth, it could stand on its own and make independent progress.

To be sure, there had already been some new work, but it was mostly like that of Walcher of Malvern, already mentioned, who prepared new astronomical tables based upon new observations but following old principles. From now on, though, European scholars were able to question the bases of the received sciences, to revise them, and to make fundamental new contributions. In astronomy, new contributions were made by Levi ben Gerson, who will be studied in the next section, by George Purbach (1423-1461, the teacher of Regiomontanus), the astronomers we have studied in earlier chapters, and still others whom there is not space to mention.

There is a characteristic of this period of European astronomy that is unfortunate from the standpoint of this work. Most of the astronomers of this period were probably more interested in understanding and transmitting the principles of their science than they were in making observations. Furthermore, modern historians of the period have been understandably more interested in the development and transmission of ideas than in the observations. Thus, of the observations that were made, relatively few have been reduced from manuscript to print and made available to the ordinary astronomer.

2.   Levi ben Gerson and His Tables

Goldstein [1974, 1974a, 1975] has made an extensive study of the works of this astronomer, and all that I have to say about him is taken from these studies. There was considerable activity in astronomy among Hebrew scholars in southern France in the 14th century, and ben Gerson† was probably the outstanding member of this group. He is perhaps best known as a philosopher, but his work on astronomy is important for its period.

Goldstein [1974, p. 19] can find no direct evidence that ben Gerson knew either Arabic or Latin, although he was well acquainted with the Syntaxis. If indeed he knew neither Arabic nor Latin, either the Syntaxis must have been translated into languages other than those I have mentioned, or ben Gerson could read the Greek directly. Further, ben

----

†Lived 1288-1344?. He is also known as Gersonides.

Gerson seems to have been ignorant of Islamic astronomy after the 11th century. In particular, he was ignorant of the tangent function, although he did go beyond Ptolemy in using the sine rather than the chord.

Nonetheless, ben Gerson made at least two important contributions to astronomy. For one thing, he rightly disliked Ptolemy's model for the motion of the moon, because it requires the apparent size of the moon to vary by a factor of almost two during the course of a single month. (See <u>Newton</u> [1977, Chapters VII and VIII] for an extensive discussion of this matter.) As a result, he developed a new model that avoided this problem while giving a reasonable description of the moon's angular position. In doing so, though, he made a serious error of principle. He argued that if the moon moved on an epicycle, it would maintain a constant orientation in space and thus present all sides to the earth.† Since the moon always presents the same side to

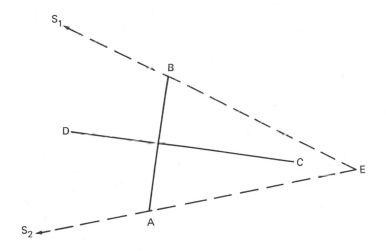

Figure IV.1 The Jacob's staff. This instrument is used to measure the angular separation between two celestial objects S_1 and S_2. A disk AB slides along the arm CD, which is held in front of the eye E, until the objects appear at opposite ends of a diameter of the disk. The accuracy obviously depends upon how well the user knows the distance from C to E.

---

†<u>Goldstein</u> [1974, p. 25] says that ben Gerson based this argument upon a principle of Aristotle. Thus ben Gerson, like most of us, could rise above his background in some matters while remaining the captive of his teaching in others. The same statement applies to Copernicus.

the earth, he argued, its model cannot contain an epicycle. Because of this error, ben Gerson constructed a model that does not contain an epicycle and as a result is highly complicated.

In contrast, an Islamic contemporary of ben Gerson named ibn ash-Shatir developed a lunar model that met the same goals by using a double epicycle for the moon [Neugebauer, 1957, p. 197], and Copernicus adopted the same model two centuries later. It is clear that ben Gerson did not influence Copernicus in this matter, and the possibility of a connection between Copernicus and ibn ash-Shatir is being studied.

A second contribution of ben Gerson is the invention of the instrument known as the Jacob's staff. This instrument, shown in Figure IV.1, is used to measure the angular separation of two celestial objects S_1 and S_2. The user holds the arm CD along his line of sight and slides the circular disk AB along it until the two objects lie at opposite ends of a diameter. If one knows the distance from C to the eye E, it is a simple matter to calculate the angle S_1ES_2. By having disks of various sizes, one can measure a wide range of angles.

Goldstein does not say how ben Gerson located the point C with respect to the eye. He does describe [Goldstein, 1974, pp. 22-23], however, some experiments that ben Gerson performed in order to determine where in the eye the point E should be taken. He concluded that it should be the center of the lens.

With regard to the invention of the Jacob's staff, Goldstein [p. 21] writes: "Some modern scholars have argued that this instrument . . . was invented prior to Levi by someone whose name was Jacob, but they fail to bring any contemporary evidence to confirm this view." . . . "Levi himself states unequivocally that he invented this instrument, and in the absence of convincing evidence to the contrary I am prepared to accept his claim." I judge from Goldstein's discussion that Levi did not call the instrument Jacob's staff, and that this name was given to it by later writers.

As we expect for such a simple hand-held instrument, the Jacob's staff is not particularly accurate. Walther, 150 years after ben Gerson, used one for about a third of his measurements of position (not counting the sun), and then abandoned it for an armillary sphere [Kremer, 1980]. A century later yet, Tycho also did some work with a Jacob's staff and quit using it when he found it subject to a scatter of about 35' [Thoren, 1973].

Goldstein gives the texts of three measurements of the position of the moon that ben Gerson made, as well as an interesting observation made during a lunar eclipse. I will take up these observations in the next section. Here, I want to look at ben Gerson's tables for the mean motion of the sun and moon.

In his Table 11† [Goldstein, 1974, p. 170], ben Gerson gives the mean position of the sun at mean noon on the last day of every year from 1301 to 1360. We should choose an epoch that is approximately in the center of ben Gerson's observations, so the end of the year 1332 is appropriate. At this epoch, the tabulated mean position is 346;1,33 degrees. Before we can use this value, we must know ben Gerson's calendrical usages and his definition of mean time.

ben Gerson used the Julian calendar, but he took the year to begin on March 1 rather than January 1. Thus the last day of the year he calls 1332 is the day we call 1333 February 28. The Julian day number of this date is 2 207 995.

He defines the equation of time as apparent time minus mean time, but his equation of time does not have an average value of zero. Instead, he defines it so that its minimum value is zero, and the average value of his equation is 4;2 "time degrees" or $16^m 8^s$. Hence, the epoch that he calls mean noon is the epoch that we call $16^m 8^s$ after mean noon.

This refers to local mean time. Orange, where ben Gerson made his observations, is at longitude 4° 48′ east (Table A.III.1), so ben Gerson's epoch is $19^m 12^s - 16^m 8^s = 3^m 4^s$ before noon, Greenwich mean time. Since the Julian day number at noon was 2 207 995, the Julian day number of ben Gerson's epoch is 2 207 994.9979. That is, according to ben Gerson's tables, this is the value of solar time when the mean longitude of the sun was 346;1,33 = 346.0258 degrees. Before we can compare this to modern tables, we must add the average value of aberration, which is 0°.0057. Thus, the geometric mean longitude of the sun at the epoch is 346°.0315.

From Equation I.24, the value of T when L_S had this value is -6.362 916 747 centuries from the epoch 2 440 400.5. The Julian day number of this epoch is 2 207 994.9658. That is, this is the ephemeris time when the mean longitude of the sun was 346°.0315. The difference ΔT is

$$\Delta T = -0.0321 \text{ days} = -2773 \text{ seconds}. \qquad (IV.1)$$

When we use Equation I.15 (Section I.4) to find y from ΔT, we get

$$y = +83.9. \qquad (IV.2)$$

We concluded in Section III.3 that we are not likely to get a useful estimate of y from any solar table newer than that of ibn Yunis [1008]. Equation IV.2 confirms this conclusion; it is obviously an unreasonable value. ben Gerson's epoch is in error by about an hour, and the declination of the sun changes by about 1′ per hour near the

---

†I do not know whether the numbering of the tables is ben Gerson's or Goldstein's.

equinoxes. Thus, ben Gerson made an error of about 1' in establishing the plane of the equator.

Table 13 of ben Gerson [Goldstein, 1974, pp. 172-173] gives the mean longitude L_M of the moon at noon on the last day of each year from 1301 to 1360, in parallel with his table of the sun. The value for the last day of 1332 is 146;57,17 = 146°.9547. Aberration changes this only to 146°.9549, so it is not necessary to make the correction for lunar aberration. For a reason that I will discuss in Section V.2, we should use the mean elongation D of the moon rather than its mean longitude. When we combine the value for L_M that we have just derived with the value for L_S that we found a moment ago, we get D = 160°.9234.

If we proceed as before, using Equation I.26 for D, we find that D = 160°.9234 when the ephemeris time was $13^m.82$ after noon, while the solar time was $3^m.07$ before noon. Thus, ΔT = +16^m.89 = 1 013^S, and

$$y = -30.9 \pm 16.3. \tag{IV.3}$$

The error estimate comes from the work of Section IX.2, where I will explain its basis.

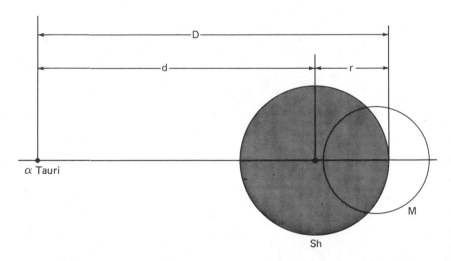

Figure IV.2 The lunar eclipse of 1335 October 3. The dark circle Sh represents the shadow (umbra) of the earth at the lunar distance. The circle M represents the moon when the magnitude was about 0.6 and increasing. At this time ben Gerson measured the distance D from the star α Tauri to the edge of the shadow that crossed the moon. By calculating the radius r of the shadow he found the distance d from the star to the center of the shadow. The moon and the shadow are approximately to scale, but the distance to α Tauri was much greater than it appears in the figure. The horizontal line represents the plane of the ecliptic. All latitudes involved were small and are ignored in the figure.

3.   ben Gerson's Observations

ben Gerson made an interesting measurement during the total lunar eclipse of 1335 October 3 [Goldstein, 1975, p. 36]. The circumstances of his measurement are shown in Figure IV.2. The dark circle Sh represents the umbra of the earth's shadow at the distance of the moon. At the time of the measurement, the moon M, moving from right to left, had about 0.6 of its diameter within the umbra, so that the right edge of the umbra struck across the bright face of the moon and was visible. ben Gerson measured the difference D in longitude between the edge of the shadow and the star α Tauri. He says that this difference, after correcting for parallax, was 43;59,23 degrees.

He uses this measurement to find the longitude of the star and from it to determine the rate of precession of the equinoxes. By calculation from his solar tables, he finds that the longitude of the sun was 196;58,34 degrees at the time of the measurement. Hence, the longitude of the center of the shadow was 16;58,34 degrees. He calculates the radius r of the umbra from eclipse theory and finds it to be 42'. Then he subtracts this from 16;58,34 to get 16;16,34 for the longitude of the edge of the shadow. Finally, he adds the measured difference in longitude, namely 43;59,23, and gets 60;15,57 degrees for the longitude of α Tauri. He compares this with earlier measurements to find the precession.†

We want to use the measurement to get the position of the sun. We must start with the measured time, which was 2;25 hours (2.4167 hours) after apparent midnight. ben Gerson does not say how he measured the time, and we must hope he measured it by astronomical sightings. At the time of the measurement, I calculate that the equation of time was +0.2150 hours, so the local mean time was 2.2017 hours, and the Greenwich mean time was 1.8817 hours. This is in the morning of 1335 October 3.

For this epoch, I find that the longitude of α Tauri was 60°.5157. From this, we subtract the measured difference in longitude, namely, 43°.9897, and get 16°.5260 for the longitude of the edge of the shadow in Figure IV.1. From modern eclipse theory, I calculate 0°.7575 for the radius r of the shadow and add it to the previous result to get 17°.2835 for the center of the shadow and 197°.2835 for the longitude of the sun. This is not quite the same thing as the apparent longitude, because aberration for this observation does not have quite the same value as for an ordinary observation, but the difference is trivial and will be ignored.

We need to discuss some optical problems involved. The earth's atmosphere "has the effect of increasing the apparent radius of the shadow by about one fiftieth" [Explanatory

---

†He decided to ignore Ptolemy's observations and to take the rate of precession as constant.

Supplement, 1977, p. 257], and I have included this effect
in my calculation of the radius r. We also have a problem
connected with the brightness of the moon.

When the bright part of the moon is seen against a dark
background, it seems larger than it really is, because of
optical effects within the eye. This arises, I believe,
either when one is seeing a bright limb of the moon against
the night sky, or when one is comparing the bright portion
of the moon with the dark portion when the moon is not full.
With regard to the bright limb, Fotheringham [1915, p. 388]
writes that a star appears occulted to the naked eye when it
is within 3' of the bright limb; he gives no reference to
any place where this finding is established. In contrast,
in Section II.4, I studied the measurements of the lunar
diameter that Tycho made with the naked eye. There I found
(Equation II.3) that the measured diameter exceeds the correct one by 1'.52 ± 0'.57. This is a correction of 0'.76
± 0'.28 to the radius, or to the apparent position of a
limb. This is about a fourth of the correction that Fotheringham applied.

To be sure, the problem of deciding where the limits of
the disk are when the moon is seen against a dark background, and the problem of deciding where a star seems to
vanish as it approaches a bright limb of the moon, may not
be the same. Further, both may be different from the problem of deciding where the edge of the umbra is as it passes
over the moon: The part of the moon just outside the umbra
is not fully illuminated, because it is in the penumbra. In
spite of these uncertainties, I will apply the correction
found in Section II.4. At least it has some data behind it,
even if they may not apply exactly to our problem.

If the bright part of the moon appears to extend 0'.76
minutes beyond its geometrical limit, we see from Figure
IV.2 that this has the effect of decreasing the radius r of
the shadow by 0'.76, and this affects the longitude found
for the sun. When we incorporate this correction, we change
the longitude to 197°.2708. This is the measured longitude
when the solar time is 1.8817 hours on 1335 October 3.

When I calculate the apparent longitude of the sun at
this solar time, using y = -20, I get 197°.1893, and when I
use y = -40, I get 197°.1969. In order to get the longitude
to be 197°.2708, we need y = -237, as we find by extrapola-
tion. This value is clearly unreasonable.

In order to give a reasonable estimate of y, the measured longitude would have to be changed by about 5'. It
takes about 2 hours for the sun to move this far. We may
safely say that ben Gerson's error in measuring time makes a
negligible contribution to the total error. It is possible
that the uncertainty in judging the edge of the umbra makes
some contribution. Further, ben Gerson says that he has
corrected the directly measured value for the parallax of
the moon, but he gives us neither the directly measured
value nor the parallax in the quotation used.

Table 23 of ben Gerson [Goldstein, 1974, p. 183] gives the parallax in the zenith distance as a function of the zenith distance itself, and from this table we find that he takes the horizontal parallax of the moon to be 53' 20", while the modern value is 57' 3". Thus ben Gerson's error in calculating the parallax may make a modest contribution to the total error.

However, all these errors are minor compared with the error that we estimate for the basic measurement. To start with, we do not know whether ben Gerson used a Jacob's staff for this measurement or not.† He was presumably well acquainted with the Syntaxis, which describes the more accurate instrument now called an armillary or armillary sphere. As we have already noted, this is the instrument that Walther chose over the Jacob's staff. Kremer [1980] finds that Walther's accuracy in measuring the longitude of a star or planet was about 9' (standard deviation). If ben Gerson used an armillary 150 years earlier, it is not likely that he did better. Since this is better than the accuracy of the Jacob's staff, let us use this estimate.

It takes the sun about $3^h.6$, or about $12\ 960^s$, to move 9' in longitude. This, then, is the standard deviation of the estimate in ΔT, and the corresponding standard deviation in y is (Equation I.15) about 400. It is clear that this observation does not give us a useful estimate of y, especially since we may be optimistic about the accuracy.

Although this observation was made during a lunar eclipse, it does not tell us anything about the position of the moon except that it was partly inside the umbra of the earth. ben Gerson merely took advantage of this opportunity to compare the positions of the sun and of α Tauri.

ben Gerson also made three direct observations of the position of the moon, which are found on pages 71, 72 and 73 of Goldstein [1974a]. The observations read, in Goldstein's translation:

> 1. "We observed the Moon with (α Scorpii) in the aforementioned year, 8;7 hours after mean noon on the 24th of June. We found its true position to be

---

†If he measured the separation of α Tauri and the edge of the shadow, and if he did not correct the separation for the difference in latitude - about 5° - he made an error of about 15' in addition to any error in the measurement itself.

Scorpio 22;19,18°."† That is, the true longitude
was 232;19,18 degrees. ben Gerson finds that this
is only 1' 5" greater than the longitude given by
his lunar model. It is clear from the context that
the year is 1333.

2. "We observed the Moon with (α Tauri) in the
aforementioned year, 6;9 hours after mean noon on
the 17th of January. We found its true position to
be Cancer 1;10°." That is, the true longitude was
91;10,0 degrees. ben Gerson finds that this disagrees with his model by only 4' 11". The year is
still 1333 for ben Gerson, who does not start his
year until March 1. It is already 1334 for us.

3. ". . . we observed the circumference (sic) of
the Moon with (α Tauri) in the year 1339 of the
Christian era, on the 11th day of December, 7;14
hours after mean noon. We found the Moon at Taurus
0;47,6°, . . ." That is, the longitude of the moon
was 30;47,6 degrees. ben Gerson finds that this
disagrees with his model by 4' 54", and he goes on
to say: "This discrepancy is probably due to the
approximation in the observation." He makes this
same remark for the preceding observation, and he
adds time keeping as a source of error for the
first observation. This probably does not mean
that he thought his time keeping was worse for the
first observation; he probably just forgot to mention it for the others.

Since ben Gerson found discrepancies between these
observations and his tables, he presumably did not use them
in constructing the tables. If this is so, the estimate of
acceleration that we derive from the observations is independent of Equation IV.3, which was derived from ben Gerson's lunar table.

For all three observations, the sun was a moderate
distance below the western horizon, the moon was about 135°
east of the sun, and the moon had not yet reached the meridian. The purpose of the observations was to study the
accuracy of ben Gerson's model when the moon was about halfway between the first quarter and full. In the first two
records, he says explicitly that the longitude is the true
longitude; that is, he has corrected the observed longitude
for parallax. He does not say "true" of the third longitude, but this is probably an oversight. He uses this
longitude as if it were the geocentric value.

ben Gerson was aware of the phenomenon of refraction,
but he probably could not deal with it quantitatively, and
he probably did not correct his observations for refraction.

---

†Two of the three manuscripts Goldstein consulted give Scorpio 22;39,18° [Goldstein, 1974a, p. 288]. Nonetheless
  Goldstein adopts 22;19,18 degrees, since this is the value
  ben Gerson uses in his calculations.

I will not do so either. In the first observation, the moon
and the reference star were separated by only a few degrees
and both were near the meridian. Hence the differential
refraction was small. In the other two observations, the
moon and the star were separated by about 30° in longitude,
but the moon was ahead in one instance and behind in the
other, and the average effect of refraction for the two
observations is almost zero. Hence, I will neglect refraction in analyzing the observations.

TABLE IV.1

THE LUNAR OBSERVATIONS OF LEVI BEN GERSON

Date	Stated time h m	Reference star	Calculated longitude of star °	Observed longitude of moon °
1333 Jun 24	20 7	α Sco	240.466	232.322
1334 Jan 17	18 9	α Tau	60.488	91.167
1339 Dec 11	19 14	α Tau	60.581	30.785

    ben Gerson specifically refers to the circumference of
the moon in the third record. Since the moon was about 3/4
full, its center was rather hard to locate. Thus it is
likely that ben Gerson measured the difference in longitude
between the reference star and the bright limb of the moon,
and then corrected this by his estimate of the radius of the
moon. Since the moon was in about the same phase for all
observations, he probably did the same thing in the other
observations, even though he does not specifically say so.

    Table IV.1 summarizes the main features of the obser-
vations. The first column gives the date of an observation
in the Julian calendar, with the year beginning on January 1
rather than March 1. The second column gives the time of
the observation as stated by ben Gerson, except that I have
added 12 hours in order to refer the observations to midnight rather than noon. In deriving the value of Greenwich
mean time that corresponds to the second column, we should
remember that ben Gerson does not use the modern definition
of mean time, and that the modern value is greater than his
by $16^m 8^s$.

    The third column of Table IV.1 lists the reference star
that ben Gerson used in finding the longitude of the moon,
and the fourth column gives the longitude of the star referred to the true equinox of date, including the annual aberration. The last column gives the longitude that ben Gerson
measured, but changed into degrees and decimal fractions.
Since ben Gerson has already corrected his measurements for

parallax, we do not know exactly what his raw values of longitude were.

We should note one point before we go any further. The reference star for the last two observations is α Tauri, which is the star whose longitude ben Gerson determined during the lunar eclipse of 1335 October 3. He found the value 60;15,57 = 60.2658 degrees, but the correct value at that epoch was 60°.5157. Thus, we should add 0°.2499 to the longitude of the moon for the last two observations in Table IV.1. ben Gerson also measured the longitude of α Scorpii [Goldstein, 1975, p. 37], but he does not give the value in any place I have seen. The error of 0°.2499 that he made in the longitude of α Tauri is enough to change the inferred value of y by about 50, versus an expected value between -10 and -20, and we must assume that his error in the position of α Scorpii is as large. Since he does not state the longitude he used for α Scorpii, it is not safe to use the first observation.

TABLE IV.2

ANALYSIS OF THE LUNAR OBSERVATIONS OF LEVI BEN GERSON

Date	Greenwich mean time h	Longitude of the moon Observed °	Calculated[a] °	Motion °/hour	y
1333 Jun 24	20.066	232.322	233.030	0.560	120.5
1334 Jan 17	18.099	91.417	91.561	0.528	10.3
1339 Dec 11	19.182	31.035	31.337	0.494	48.0

[a]Using $\dot{n}_M = -28$ and $y = -20$.

Although I will not use the first observation, I will analyze it so that the reader may use it if he wishes.

The analysis of the observations is summarized in Table IV.2. The first two columns give the date and the equivalent, in Greenwich mean time, of the stated hour of the observation. The third column gives the observed longitude of the moon. This value is taken from Table IV.1 for the first observation, in which the reference star was α Scorpii. For the other observations, in which the reference star was α Tauri, I have added 0°.2499 to ben Gerson's value and then rounded to three decimal places.

I next calculated the longitude of the moon, and its hourly motion in longitude, for the value of time given in the second column, using the accelerations $\dot{n}_M = -28$ (Equa-

tion I.23) and y = -20. These values appear in the fourth
and fifth columns of the table. In using these, consider
the first observation as an example. At the time when the
observed longitude was 232°.322, the calculated longitude
was 233°.030. The calculated longitude would equal the
observed longitude at a time 1.2636 hours earlier than the
time of the observation. Thus, if we use CT to denote the
time just calculated and ST to denote the tabulated value of
Greenwich mean time, we have

$$CT - ST = -1^h.2636 = -4\ 549^S.$$

If I had used y = 0 when I calculated the values in the
fourth column of Table IV.2,† CT would be the same thing as
ephemeris time, and -4 549 seconds would be the difference
ΔT between ephemeris time and solar time. However, I used y
= -20 in the calculations, and thus I took as my starting
estimate that ΔT has the value given by Equation I.15 with y
= -20; this value is +643 seconds. The value -4 549^S is the
amount by which we must change the starting estimate. We
see from Equation I.15 that, if we change ΔT by -4 549^S for
an observation at this epoch, we must change y by +140.5.
Hence, the inferred value of y, which appears in the last
column of Table IV.2, is +120.5.

We decided a moment ago not to use this value, which is
from the first line in Table IV.2, and to use only the last
two lines. The average value of y from these lines is
+29.1, and it only remains to attach a standard deviation to
this estimate.

Let us take the standard deviation of a measurement of
longitude to be 9', as we did before. It takes the moon
about 984 seconds to move this far, and the corresponding
change in y for observations of this age is 30.4. Since two
observations were used in forming the estimate, the appropriate standard deviation is 30.4/√2 = 21.5. Thus we get

$$y = +29.1 \pm 21.5 \tag{IV.4}$$

from ben Gerson's measurements of the moon. The epoch for
this value of y is 1336.

We can combine this with Equation IV.3 to get a single
estimate from the work of ben Gerson. The result is

$$y = -1.9 \pm 13.1. \tag{IV.5}$$

We may still take the epoch to be 1336.

---

†It is legitimate to ask why I do not use y = 0 in the calculations. The answer is that the calculated circumstances of an observation are not linear functions of y, but
it is convenient to use linear methods in the analysis.
This means that I should use a value of y that is not too
far from the correct one.

4. The Alphonsine Tables

Alphonso X, called the Learned, was born about 1226, and was king of Leon and Castile from 1252 to 1284.† Alphonso has some standing as a poet, he encouraged and sponsored the study of history and the writing of a national history, and he sponsored the preparation of the astronomical tables known as the Alphonsine tables.

The Alphonsine tables had a long run of popularity. After surviving in numerous manuscript copies for more than two centuries, they were first printed in Venice in 1483 and last printed in Madrid in 1641 [Dreyer, 1920]. There were about a dozen printed editions, and the total number of manuscripts is probably unknown. Dreyer examined twelve manuscripts in Oxford alone.

According to the standard legend, Alphonso assembled a congress of fifty Arabic, Hebrew, and Christian astronomers and instructed them to prepare new astronomical tables, which were completed in 1252. Since the preparation of the tables, by instruction, entailed a considerable period of observation, and since Alphonso only came to the throne in 1252, the legend should attribute the tables to Alphonso's predecessor in order to be consistent.

Actually, as Dreyer [1920] emphasizes, there is documentary evidence about the preparation of the tables which shows that the preceding is legend. In volume 4 of Rico [1867], there is an introduction to the Alphonsine tables that is clearly the original introduction. This introduction shows that the entire work was carried out in the years 1263 - 1272 by two Hebrew astronomers whose names may be rendered as Jehuda ben Mose and Isaac ibn Sid. From the same source, we learn the reason for the new tables: The Toledan tables, which had been prepared by al-Zarkala in Toledo in about 1080, now differed significantly from observation. Hence Alphonso instructed ben Mose and ibn Sid to make new observations in Toledo, and to use them as the basis for new tables.

The tables in all the printed editions, and in many of the later manuscripts, differ drastically in form from the

---

†As a rule, I try to use the local spelling of a European name rather than the English spelling. With Alphonso, there is no recognized Spanish spelling. It is spelled Alfonso in Castilian and Alonso in Leonese.

tables in the earliest manuscripts,† although they give the
same basic information, For example, the printed tables
give the mean longitudes of all the bodies for the beginning
of the Christian era. The user must then figure the number
of days (and fractions thereof) from this era to the day and
time he wants to use, and he must express this time in full
sexagesimal notation. That is, he must write the time, in
units of days, in the form

$$. . + \underline{a}_3 \times 60^3 + \underline{a}_2 \times 60^2 + \underline{a}_1 \times 60 + \underline{a}_0 + \underline{a}_{-1} \times 60^{-1} + . .,$$

instead of using years, days, hours, and so on. The tables
give the changes in mean longitude that correspond to the
coefficients $\underline{a}_3$, $\underline{a}_2$, and so on. Finally, to get a mean
longitude, the user adds all of these changes to the value
at the beginning of the era.

Actually this type of table is easy to use once the
user has found the coefficients $\underline{a}_3$, $\underline{a}_2$, and so on. To give
an example in decimal notation, suppose that the mean motion
of the moon in longitude is $\underline{n}$ degrees per day, and that we
have a table of $1 \times \underline{n}$, $2 \times \underline{n}$, and so on to $9 \times \underline{n}$. If we
want the change in mean longitude for 267 days, say, we
first look in the table for the change in 7 days. Then we
look in the table for the change in 6 days and shift the
decimal point one position. Next we look up the change in 2
days and shift the decimal point two positions. Then we add
up these values and add the result to the initial value.

The earliest manuscripts of the tables of mean motion
are quite different. To start with, they give the actual
mean longitude at noon on December 31 of every year that is
a multiple of 20 years. Then they give the change, modulo
360°, for every single year, paying due attention to leap
years. Following this, they give the changes from the end
of December up to the end of each following calendar month.
Finally, they give the changes for integer days and integer
hours. Tables for integer minutes are not necessary, since
we just take the tables for hours and shift the sexagesimal
point. This form of the table is easier for the user, since
he needs only the date in the Julian calendar, plus the
hours and fractions thereof, in order to use them. On the
other hand, this form requires larger tables.

Dreyer [1920, p. 244] makes a peculiar remark at this
point. He says that the sexagesimal (printed) form of the
tables was probably made in slavish imitation of Ptolemy.
It seems to me that the manuscript form is the one that
imitates Ptolemy. In fact, Ptolemy's tables of mean motion
have the same basic structure as the manuscript form. The

---

†Immediately after the introduction to the tables, Rico
prints what he calls "Numerical Fragments of the Alphonsine
Tables", which at first look like still a third form of the
tables. Actually they are tables from what Dreyer calls a
"perpetual almanac"; as Dreyer says: "It is difficult to
understand how Rico could imagine that this almanac had
anything to do with the Alfonsine Tables."

only significant differences are that Ptolemy uses intervals of 18 years rather than 20, and that he uses the Egyptian year and month rather than the Julian ones. The basic form of the printed tables is Babylonian rather than Ptolemaic, but how did the printers know about Babylonian astronomy?

I do not understand the theoretical structure back of the tables. The point of difficulty is the precession of the equinoxes. As I have already mentioned in connection with Copernicus (Section II.7), astronomers who tried to use Ptolemy's "observations" as well as authentic ones were forced to conclude that the motion of the equinoxes is not uniform,[†] and this led them to the theory that became known as the trepidation of the equinoxes. Many medieval astronomers, such as ben Gerson and ibn Yunis, rejected the idea of trepidation. Unfortunately, many others, including Copernicus and the makers of the Alphonsine tables, accepted the idea and thereby considerably lessened the value of their work.

ben Mose and ibn Sid assumed that the motion of the equinox is made up of two parts. One part constitutes a uniform precession at the rate of 26".45 per year. The trepidant part has the form $9°$ sin $[(360°)(Y/7000)]$, where Y is in years. That is, it is a sine wave with an amplitude of $9°$ and a period of 7000 years. Y is measured from the epoch 15 May 18.[‡]

At the end of the year 1260 (1260 December 31), Y is about 1245.6 years, and the trepidant part equals $8°.093$ while the uniform part (from 15 May 18) equals $9°.152$, for a total motion of $17°.245$. This compares favorably with the correct amount of precession over this interval, which amounts to $17°.296$. However, at the epoch 1260, the rate of precession, coming from both the uniform and trepidant parts, is only about 38" per year instead of the correct amount of about 50".

Now the motion of the equinoxes should not affect the motions of the sun and moon among the stars. That is, the truly sidereal year and month should be independent of the motion of the equinoxes. Thus, if there is trepidation in the motion of the equinoxes, there should also be trepidation in the tropical year and month. However, the lengths

---

[†] I am neglecting the small nutations and the small secular acceleration of the equinoxes in this discussion.

[‡] There are two points of interest about these parameters. First, the epoch 15 May 18 is about midway between Hipparchus and Ptolemy. Second, the length of the year in the Alphonsine tables is 365^d 5^h 49^m 16^S, which is only 30 seconds longer than the accurate value. It is 10^m 44^S less than the Julian year, and the sun travels 26".45 during this time [Dreyer, 1920, p. 250]. It looks as if ben Mose and ibn Sid used a little "numerology" in deriving their parameters of precession; the Julian year should play no fundamental part in astronomy.

of the tropical year and month used in the tables are strictly constant, the year at 365.242 546 days and the month at 27.321 586 days.† These are lengths of the year and month that go with accurate rates of precession.

Dreyer [1920, p. 250] says that the instructions for using the Alphonsine tables point out that there should be trepidations in the length of the year and month and explain how to find them. Dreyer goes on to write: "But this was apparently considered a needless refinement both in the Oxford redaction‡ and in the printed tables, in both of which this slight irregularity is ignored. This is very excusable, since the periodic term in the fourteenth and fifteenth centuries diminished at the rate of only 9″ or 10″ in twenty years, which in those days was a negligible quantity."

Actually the periodic term was increasing rather than decreasing in the fourteenth and fifteenth centuries, and at a rate of about 4′ (240″) in 20 years. This was definitely not negligible even then, but the tables show no evidence of this effect. To be specific, when the tables were prepared, say about the year 1260, the total rate of precession from both the uniform term and the trepidation was about 38″ per year, about 12″ less than the correct value. The sun takes about 288 seconds to go 12″ and the moon takes about 22 seconds. Hence the tropical year in the time of ben Mose and ibn Sid should have been too long by about 288 seconds and the tropical month by about 22 seconds, but the actual errors are about 30 seconds and 0.3 seconds, respectively, as we have seen.

Thus it seems that, at least for the sun and moon, the makers of the Alphonsine tables let the trepidation affect only the value of the anomaly but not the mean motion in longitude.

Further, although ben Mose and ibn Sid let their work be affected by Ptolemy when they adopted the trepidation of the equinoxes, they did not go so far as to use his claimed solar observations in finding the length of the year. In the observations that Ptolemy claims to have made, the sun reaches any particular point in its orbit slightly more than a day late. Thus, if a later astronomer used his solar observations to find the year, he would get a value considerably too short instead of being slightly too long. In fact, so far as I can make out in a brief investigation, ben Mose and ibn Sid probably found the year by combining their observations with the summer solstice observed by Aristarchus in −279.

---

†The year, as we just noted, is too long by 30 seconds, and the month is too long by about 0.3 seconds.

‡This refers to a manuscript form of the tables found in Oxford.

Most copies of the tables that preserve the original
form begin in 1340 or even later, with only one copy known
to Dreyer that begins in 1320.† It appears that copiers of
the tables generally started only with their own times and
dropped out the earlier portions. Dreyer [1920, pp. 255-
260] prints the tables that deal with mean motions and the
equinox from a copy that runs from 1320 to 1600. In this
copy, the meridian has been changed from that of Toledo to
that of Oxford, which is taken to be 16 minutes (4°) east of
Toledo. Actually, Oxford is not this far east of Toledo,
but Greenwich is almost exactly so. Thus, by a lucky coincidence, we can apply the tables to Greenwich without
change.

Because of the problems about the equinox, I do not
think it would be safe to use the individual tables of the
sun and moon, even if the solar table could be considered
accurate enough. However, whatever ben Mose and ibn Sid did
about the equinox, they should have done the same thing with
both the moon and the sun, and the uncertainty should cancel
if we subtract a solar position from a lunar one. In other
words, we can construct a table of the lunar elongation from
the Alphonsine tables and use it in our study.

TABLE IV.3

SOME MEAN POSITIONS FROM THE ALPHONSINE
TABLES FOR NOON ON 1261 JANUARY 0

Quantity	Value from the tables °	Value from modern tables °
Mean longitude, sun	287.596	287.355
Mean longitude, moon	257.116	256.583
Mean elongation, moon	329.520	329.228

We should use the table at an epoch close to the time
when ben Mose and ibn Sid made their observations, and the
closest multiple of 20 years to this time is 1260. Since
the table gives the mean positions of the sun and moon for
1320 and 1380, we find the change in mean position in 60
years by combining these values, and thence we find the
positions for 1260. For this epoch, the relevant quantities

---

†The years are not always multiples of 20 in the copies.
Dreyer found one copy with values given for 1348, 1368, and
so on to 1628.

are listed in Table IV.3. The epoch is noon, Greenwich time, on 1260 December 31, which is the same as 1261 January 0. The values calculated from modern tables have been corrected for aberration.

The time base for the last column is ephemeris time, with $\dot{n}_M = -28$ and $y = 0$. From Table IV.3, we need to calculate the ephemeris time when the mean elongation was 329°.520. Since the elongation was 329°.228 when the ephemeris time was noon, and since the elongation increases by 12°.191 per hour, the elongation reached 329°.520 at 2 069 seconds after noon. That is, ΔT in Equation I.15 equals 2 069 seconds, and $y = -46.9$. We need to attach a standard deviation to this estimate.

In the preceding section, I used 9' as the standard deviation of a single observation. At the time of the Alphonsine tables, I think we should increase this to 15', say. In constructing a table of mean positions, I think most ancient and medieval astronomers combined one of their own observations with an observation that was as old as they could find. This means that the error in the resulting table in the time of the Alphonsine astronomers is the same as the error they made in their own observation. If we take 15' as the standard deviation in the value of the elongation at the beginning of 1261, we get 40.0 as the standard deviation of y. Hence we get

$$y = -46.9 \pm 40.0 \qquad\qquad (IV.6)$$

from the Alphonsine tables.

Goldstein† [1982] derived a value of y from the solar table alone, using a later printed form of the tables that is based upon the sexagesimal expression of the date. He discarded the resulting value because it was so discordant with other results. He did not try using the table for the elongation.

5.   The Solar Eclipse of 1221 May 23

In the spring of 1221, when Genghis Khan ruled over much of Asia, a Chinese party made an expedition across central Asia, and one of the members of the expedition kept a log of the happenings. Wylie [1897] gives this translation of a certain passage from the log: " . . at noon, an

---

†This is S. J. Goldstein, Jr., whereas the Goldstein I have cited earlier is B. R. Goldstein.

eclipse† of the sun happened, while we were on the southern
bank of the (Kerulen) river. It was so dark that the stars
could be seen, but soon it brightened up again." This is
the record that I designate as 1221 May 23a C in volume 1
and in Appendix I.

At two places which the expedition reached subsequently
they compared their record with local records of the
eclipse. One of the places cannot be identified, but the
other is Samarkand. There the magnitude of the eclipse was
recorded as six-tenths.

In earlier times, astronomers often gave the magnitude,
not as the fraction of the diameter that was covered during
an eclipse, but as the fraction of the area that was covered. Figure A.II.3 in Appendix II gives the relation between the two measurements of magnitude for a solar eclipse.
The text gives no clue to the meaning of magnitude that is
used in this record. Since the reference of magnitude to
the area rather than to the diameter seems to be the earlier
usage in regions of the Near East, I assume that the area is
used here, even though Far Eastern practice need not have
been the same as Near Eastern. This is the same choice I
made earlier [Newton, 1970, p. 145].

From Figure A.II.3, we see that a magnitude of 0.6 of
the area corresponds to a magnitude of 0.68 of the diameter.
In assigning a standard deviation to this value in the place
just cited, I originally made the naive assumption that the
precision in stating a measurement is related to the observer's estimate of its accuracy. On this basis, I at first
took the standard deviation of the magnitude to be 0.04.
Later [p. 244], on the basis of the analyzed results of many
records, I raised the standard deviation to 0.07. I use
this value here. We take the coordinates of Samarkand from
Appendix III.

The magnitude of an eclipse has the same value at
points that are symmetrically placed with respect to the
center line. In dealing with an eclipse that may have been
central, the magnitude is a poor quantity to deal with because of this double-valued property. For this reason, I
use the coordinate η in volume 1 and in Appendix I; η has
the property that it is negative on one side of the center
line and positive on the other. Thus, it has a zero instead
of a maximum on the center line.

However, for an observation in which the magnitude was
measured, and for which the magnitude was far from totality,
our calculations need never deal with a point close to the

---

†In the translation of this passage that is used by Muller
and Stephenson [1975] and by Muller [1975], the eclipse is
described as total, and the writers just cited say that the
term used to indicate totality is chi or jih. I studied
this term carefully in volume 1 and showed that it did not
mean totality "at any stage in history" from which we have
useable records of eclipses.

center line. In such a case, we may as well use the magnitude, which I denote by $\underline{m}$. Using the standard values $\underline{\dot{n}}_M$ = -28 and $\underline{y}$ = -20, I calculate $\underline{m}$ = 0.628 for the eclipse of 1221 May 23 as observed at Samarkand, and I calculate that $\partial \underline{m}/\partial \underline{y}$ = +0.0028. I also find that $\partial \underline{m}/\partial \underline{\dot{n}}_M$ = -0.0041.

With these values, assuming as usual that $\underline{\dot{n}}_M$ = -28, we find

$$\underline{y} = -1.4 \pm 25.0. \tag{IV.7}$$

6. The Observations of Walcher of Malvern

The name Malvern is applied collectively to a number of neighboring towns and villages that lie about 50 kilometers southwest of Birmingham. The district once contained a number of important ecclesiastical establishments, including a Benedictine priory, and the individual known as Walcher of Malvern was attached to one of these.

What little I know about Walcher comes from Haskins [1924, pp. 114-115]. About the year 1110, he prepared some lunar tables that run for 76 years and that end in 1112. He based them in part upon an eclipse of the moon that he observed in 1092. Of course, he must have used some other information, but I do not know what it was. We note that 76 years is an important cycle of the lunar motion. The tables are preserved in a manuscript that Haskins cites.

Walcher has left us records of two lunar eclipses, and, in fact, of three if my reading is correct.

1091 October 30. Walcher saw this eclipse at an undesignated place that was 1 $\frac{1}{2}$ days journey east of Rome. At the time of the eclipse, the moon was in the west and the time was shortly before dawn. When he returned to England, he compared records with a brother monk and concluded that there was a large time difference between Italy and England. In Italy, the eclipse had been a little before dawn with "the moon verging toward its setting." In contrast, "here in our island of England", the time was before midnight with the moon still ascending. These data imply that Italy is about a quarter of the way around the world from England. Could Walcher really have thought that he had travelled that far?

The Italian part of this record makes sense but the English part does not. According to Oppolzer [1887], the middle time of the eclipse of 1091 October 30 was $5^h\ 27^m$, Greenwich time, which was about $6^h\ 20^m$ at a point slightly east of Rome. In 1091, the Julian calendar was 6 days ahead of the Gregorian calendar, so the autumnal equinox was on about September 15, and the eclipse was about 45 days after the equinox. Sunrise at the latitude of Rome was about $6^h\ 40^m$, so the eclipse should have been somewhat past maximum when the moon set.

Matters were not far different in Malvern. The half-duration of the eclipse was about 78 minutes, so it should have begun about $4^h 9^m$ (neglecting the time difference between Malvern and Greenwich) and it should have been over about $6^h 45^m$, just about at sunrise and moonset. This could not have been recorded as being before midnight.

I believe that Walcher's brother monk was describing the eclipse of 1091 May 5 rather than that of 1091 October 30. Both eclipses had magnitudes slightly greater than 0.5, and both had half-durations of about 78 minutes. It may be that the eclipse in May was visible in Malvern while the one in October was not visible because of the weather. Thus, when Walcher later asked his brother if he had seen the eclipse in 1091 when the moon was about half eclipsed, the brother would have thought he meant the eclipse of May.

Whether this explanation is correct or not, we cannot use this record. Apparently no attempt was made to measure the time of the eclipse in England. Walcher says that he had with him in Italy no instrument to measure the time with. The vague statements of time we have are not accurate enough to use in this work.

1092 October 18. Walcher makes no statement about travelling in connection with this observation, so we do not know whether he was back in Malvern or still in Italy. I will calculate the circumstances for both Malvern and Rome, on the assumption that he was in or near Rome if he was in Italy. For this eclipse, Walcher measured the time when totality began and found it to be when 11;3 hours had been completed. Walcher says that he measured the time by means of an astrolabe. The problem with this record is to interpret the statement of the time.

In western civilization in 1091, and for many centuries before and for a few centuries thereafter, an hour in ordinary usage did not mean 1/24 of a day. Instead, the interval between sunrise and sunset was divided into 12 equal parts called hours of the day, or some such term. Similarly, the interval between sunset and sunrise was divided into 12 equal parts called hours of the night. In this usage, a statement of the hour was referred to the preceding sunrise or sunset rather than to noon or midnight. This suggests that "11;3" refers to hours of the night measured from the preceding sunset.

Our first task in analyzing the record is to convert the statement of time into Greenwich mean time, taking Malvern and Rome as the possible places of observation. The towns and villages to which the name Malvern is attached seem to have Great Malvern as their center, and Great Malvern also seems to be the site of the earliest ecclesiastical establishment in the region. Hence, I take it as the specific location meant by Malvern; any error in this assumption is inconsequential. The coordinates of Great Malvern are given in Appendix III.

The eclipse came between sunset on 1092 October 17 and sunrise on October 18. At Great Malvern, the sunset in

question came at 16.961 hours, Greenwich mean time, and the
sunrise came at 6.864 hours, also in Greenwich mean time.
Thus 1 hour of the night equalled 1.1585 of our hours. Now
11;3 hours of the night means 11.05 of them in decimal notation, so the time recorded by Walcher was 12.802 ordinary
hours after the preceding sunset. This was 5.763 hours,
Greenwich mean time, on 1092 October 18.

At Rome, the sunset was at 16.291 hours and the sunrise
was at 5.551 hours, both in Greenwich mean time. Hence, the
time stated by Walcher was 4.502 hours, Greenwich time.

I calculate the following for the Greenwich hour $\underline{h}$ at
which the eclipse became total:

$$\underline{h} = 4.190 \text{ hours if } \underline{\dot{n}}_M = -28 \text{ and if } \underline{y} = -20,$$

$$\partial \underline{h}/\partial \underline{y} = +0.0214, \qquad\qquad\qquad (IV.8)$$

$$\partial \underline{h}/\partial \underline{\dot{n}}_M = -0.0227.$$

As usual, I assume that the value of $\underline{\dot{n}}_M$ is correct, and I
infer only the value of $\underline{y}$. Before finding the value of $\underline{y}$,
we need to discuss the accuracy of the measured time.

In Section IX.2, I analyze a large body of times associated with lunar eclipses that were measured by Islamic
astronomers at a slightly earlier stage in history. The
standard deviation of these times comes out to be 0.23
hours. Although Walcher is later, it is doubtful that he
achieved greater accuracy, since Europeans were relative
newcomers to astronomy. Hence, I take 900 seconds as the
standard deviation of the measured time. For the year 1092,
this yields 11.7 as the standard deviation of $\underline{y}$.

If the observation were made in Malvern, the observed
hour was 1.573 hours after the calculated one. From the
second of Equations IV.8, then, we must add 73.5 to the
assumed value of $\underline{y}$, giving us $\underline{y} = 53.5 \pm 11.7$. We can be
sure that the correct value is negative, so this estimate is
in error by at least five standard deviations. We may
safely discard this solution.

If the observation was made in Rome, on the other hand,
we get

$$\underline{y} = -5.4 \pm 11.7. \qquad\qquad\qquad (IV.9)$$

This is a plausible solution, and I adopt it. Equation IV.9
tells us that Walcher was still in Italy on 1092 October 18.

7.   The Solar Eclipse of 792 November 19

The Chinese astronomical records that deal with the
Tarng dynasty (618 - 907) list the times of seven solar
eclipses. More specifically, they give a time interval,
that ranges from about 15 minutes to 1 hour, within which
the maximum eclipse occurred. These times occur in a docu-

ment that was completed in 1370, and they do not appear in any earlier known source. It is likely that these times were calculated by Chinese astronomers not long before 1370 as a test of the accuracy of their current calendar [Cohen and Newton, in press], although there are some difficulties with this hypothesis. Certainly it is not safe to use these times as observed data.

The contemporaneous records also refer to the magnitudes of eight eclipses in terms which indicate that the references are the result of observation. In all cases but one, however, the records give only qualitative information about the magnitude, such as a statement that the eclipse was almost "complete". The one eclipse for which we have a quantitative measure of magnitude is the eclipse of 792 November 19. I take the translation of this record from Cohen and Newton.

The record of this eclipse reads (note 99 to Table 3 of Cohen and Newton): "Previous to this day [the pertinent official] submitted a memorial that according to the calendar numeration there will be an eclipse [with a magnitude] of eight parts. Now we retract (?) the eclipse [magnitude] to three parts. The reckoning is decreased by a 'strong half' (?)." There are some questions about the words that are followed by a question mark, but only the last one concerns us.

The Chinese unit for the magnitude of an eclipse was [Yabuuchi, 1944, p. 18] one part in fifteen, but we do not know whether this referred to the fraction of the diameter or of the area that was eclipsed. Estimating the fraction of the area was the more primitive practice in the regions near the eastern end of the Mediterranean, but this does not necessarily make it the early Chinese practice. Nonetheless, since it is necessary to make a decision, I assume that the observed magnitude of 3/15 refers to the area. This is the same choice that I made for early Chinese records in a previous work [Newton, 1970, p. 145], but I emphasize again that there is no textual basis for it.

The record says that the predicted magnitude was 8/15 but that the observed magnitude was only 3/15. The difference is 5/15 = 1/3, but the record seems to refer to a difference that almost equals 1/2. This suggests that there is a scribal error in either the predicted or the observed magnitude. However, the reading of the difference is not certain, so I will take the observed magnitude to be 3/15 = 0.2 of the area. From Figure A.II.3 in Appendix II, we see that this is about 0.31 of the diameter.

We do not know where the observation was made, and we can only put it somewhere in what I called "heartland China" in volume 1 (p. 141). Luckily, the uncertainty in the position of the observer is not particularly important for an observation this old. The corners of heartland China are given in Appendix III.

In an earlier work [Newton, 1970, p. 144], I listed the magnitudes of two solar and three lunar eclipses that are

found in Chinese records between 585 January 21 and 592
August 28, and I used them in studying the accelerations.†
At the time I did this earlier work, I had not realized the
extent to which Chinese astronomers, even as early as -200,
were calculating the circumstances of eclipses and entering
the calculated circumstances into the archives. Relatively
few records of eclipses in the Chinese archives can safely
be taken as the result of observation, and I will not use
the records mentioned at the beginning of this paragraph.
However, it is safe to use the record of 792 November 19,
since it explicitly compares the observed magnitude with the
predicted one.

For convenience, I will analyze this observation in
Section IX.3, along with some Islamic observations of a
slightly later period.

8.   The Lunar Eclipse of 755 November 23

The eclipsed moon must frequently be close to a bright
star or planet, but I can think of only two records before
the modern period which actually record such an event. One
is the record of the lunar eclipse of 1335 October 3 made by
the professional astronomer Levi ben Gerson, which I studied
in Section IV.3. The other is a record found in a monastic
chronicle [Simeon of Durham, ca. 1129]. Simeon, drawing
upon the archives of the monastery at Durham, wrote the following about the total lunar eclipse of 755 November 23:
"And remarkably indeed a bright star east of the moon passed
through it, and after the illumination it was west of the
moon by as much space as it had been east before the
eclipse."

When I first analyzed this report [Newton, 1972,
Section XVII.1], I looked for the star that was closest to
the center of the moon at the center of the eclipse, and
found that this star was 114 Tauri, a star of magnitude 4.8.
Compared to the eclipsed moon, such a star might conceivably
be called bright, but nonetheless this identification seemed
unlikely. Even so, it did not occur to me to see if the
"star" might be a planet. However, almost immediately after
the appearance of my study, Ashbrook [1972] pointed out that
the "star" was in fact Jupiter.

If Jupiter was as far west of the moon after the
eclipse as it had been east before the eclipse, the moon and
Jupiter were in conjunction with each other, and both were
in opposition to the sun, at the center of the eclipse. In
the context of this observation, parallax is negligible for
Jupiter but it is important for the moon, and we must use
the position of the moon as seen at Durham. At the center
of the eclipse, I calculate that the apparent longitude of
the moon was 65°.48 and that the longitude of Jupiter was

---

†I also listed the magnitude of the lunar eclipse of 593
August 17, but I did not use it because of suspicious
circumstances.

66°.12. Since the radius of the moon is at most 0°.56, the
moon had not even begun to occult Jupiter at the time of
greatest eclipse. The error in the observation is about
0°.64.

Of course, we cannot expect a monk of the 8th century
to make astronomical observations with the accuracy we expect of a trained astronomer. Still, we can wish that he
had been a little more careful in his description of the
eclipse-occultation. We regretfully conclude that we cannot
use the interesting observation of 755 November 23 in this
work.

CHAPTER V

ISLAMIC OBSERVATIONS

1. The Sources

Islamic armies "erupted" out of Arabia in the year 634 and rapidly overran their Asian neighbors before they turned their attention to Africa. By about 640 they had taken Alexandria and most of Egypt from the Byzantine empire. When they took Alexandria, it has been widely charged that they used all the books in its famous library as fuel for the public baths. While many peoples have shown themselves capable of the kind of religious bigotry described in this story, there is little chance that this particular instance of it really occurred. See Appendix IV.

From Egypt, the Muslims spread rapidly across the rest of northern Africa, across Spain, and into France, where their advance in the west was finally halted by Charles the Hammer in a battle in 732. The extent of their greatest advance is shown by the location of this battle, which was somewhere in the region between Tours and Poitiers. Even after this battle, the Muslims clung to some French territory until about 760. At a later date, the Muslims also conquered all of Sicily.

At least parts of Spain and Sicily remained under Muslim domination for many centuries, with important consequences for the civilization of western Europe. The Muslims came into contact with ancient culture in its conquests around the eastern end of the Mediterranean and brought some elements of this culture to Spain and Sicily, along with contributions of their own.† These areas, as I noted in Section IV.1, were probably the most important ones for the transfer of ancient learning to Europe.

It seems to have taken some time for the Muslims to become interested in astronomy; the first Muslim work in astronomy I know of that can be dated comes about 750. Early Muslim astronomy was derived from both Indian and Greek astronomy. Since ancient Indian astronomy did not reach the level attained by the Greeks, early Muslim astronomy is variable in quality. After a fairly short period, however, the Muslims adopted the more sophisticated Greek astronomy but, as I have said in a note, they did not go far beyond it. Neither did the Europeans until the time of Tycho and Kepler.

---

†In astronomy, the Muslims developed an improved model for the motion of the moon, but otherwise they went little beyond the Greeks except for the making of new observations and the consequent improvement of parameters. In algebra and trigonometry, however, they went far beyond the Greeks.

After a late start, Muslim astronomy grew rapidly and flourished for a long time. Kennedy [1956] lists 109 Muslim works of the type called a zij dated between about 750 and about 1450; a zij in Kennedy's definition is a document that contains a "more or less complete set" of astronomical tables that may or may not contain an explanation of the underlying theory. Unfortunately for our purposes, few Muslim works on astronomy, whether they are zijes or not, have been published in the form of a translation into a European language. Kennedy [1956, p. 128] remarks that only two of the zijes in his list have been published,† and one work that I know of has been published since [al-Biruni, 1025]. Thus, of all the Muslim sources that exist, I am limited to the study of four in this volume, namely al-Biruni, ibn Yunis, al-Battani, and al-Khwarizmi. I will discuss these sources in the rest of this section, starting with the latest and going back in time to the earliest.†

al-Biruni was born [Kennedy, 1970] on 973 September 4 in the region formerly known as Khwarizm,‡ and he died in Ghazni, Afghanistan sometime after 1050. His full name is Abu-Rayhan Muhammad bin (or ibn) Ahmad al-Biruni. It was common then to call a Muslim after the place of his birth, and Kennedy says that he was born in a suburb (birun) of the city of Kath. The town now known as Biruni was named for the scholar, not the scholar for the town.

---

†These two are the works by al-Khwarizmi and al-Battani, whom I will discuss later, which have been published in Latin translation. Apparently Kennedy is referring only to zijes in their complete form, because portions of the zij of ibn Yunis were published in Arabic, with a parallel translation into French, in 1804. See the reference to ibn Yunis [1008].

†There is also a work by Thabit bin Qurra that was written about 880 and that was translated into Latin by Gerard of Cremona about 1170 under the title De Anno Solis. This work deals only with the motion of the sun and does not qualify as a zij under Kennedy's definition. I used the solar data from De Anno Solis in an earlier work [Newton, 1976, Chapter VII]. There is an edition of Gerard's translation by F. J. Carmody, with notes and commentary, in The Astronomical Works of Thabit b. Qurra, University of California Press, Berkeley, California, 1960.

‡Kennedy says that this is now the region of the Kara-Kalpak Autonomous Soviet Socialist Republic, which borders on the Aral Sea. This contradicts both Toomer [1973] and my copy of the Geographical Dictionary [Webster's, 1949], who both agree that the region is still known as Khoresm and that it is in the Uzbek Soviet Socialistic Republic. However, Khoresm and Kara-Kalpak are close together and both border on the Aral Sea, so the difference is not geographically important, although it may be politically important. Most of this discussion of al-Biruni's life is taken from Kennedy [1970].

We do not know anything about the ancestry or early life of al-Biruni, but we do know that he was deeply interested in astronomy by what we would call college age. At the age of seventeen, he found the latitude of Kath by measuring the meridian altitude of the sun. A few years later [al-Biruni, 1025, p. 77] he made a circle 15 cubits (about 7 meters) in diameter, with which he planned to make a systematic series of observations over a period of two years in Khwarizm. However, he had had time to make only one observation, that of the summer solstice of 995, when he was forced by warfare to flee his native land. From this time on, his life was almost continually embroiled with the politics and warfare of his time and place. A few years after he fled, he says that "I was permitted by the Lord of Time to go back home, but I was compelled to participate in worldly affairs, which excited the envy of fools, but made the wise pity me." I think we may suspect that this sentence does not convey his true feeling toward involvement in "worldly affairs". I refer the reader to Kennedy [1970] for further discussion of al-Biruni's life.

Kennedy [1956, pp. 157ff) gives considerable detail about the zij of al-Biruni which, I believe, has been neither published nor translated in spite of its historical importance. His only published work, so far as I know, is his work on geodesy [al-Biruni, 1025]. In this work, al-Biruni describes a number of methods for determining the coordinates of a point, and he also describes how to calculate the azimuth of the great circle that connects any other point with Mecca; knowledge of this azimuth allows a Muslim to face in the correct direction when he prays.

Most of the observations given in al-Biruni's work are solar ones, and I have analyzed them elsewhere [Newton, 1972a]. They give us

$$y = -26.5 \pm 5.8 \qquad\qquad (V.1)$$

in the usual units. Use of the DE102 ephemeris instead of Newcomb's theory might change the value slightly.

In addition to the solar observations, al-Biruni gives observations of one lunar and one solar eclipse. I will describe these observations in Sections V.3 and V.4.

Next we turn to ibn Yunis, whom King [1976]† calls "one of the greatest astronomers of medieval Islam." His full name is Abul-Hasan Ali ibn Abd al-Rahman ibn Ahmad ibn Yunis al-Sadafi. Unlike the other Muslim astronomers with whom we have to deal here, ibn Yunis is known by a patronymic instead of by his "geographical" name al-Sadafi. King describes him as coming from a respected family. His great-grandfather Yunis was well-known in legal circles and his

---

†King gives the name as ibn Yunus, but I believe that the spelling ibn Yunis is more common in English writing. I have taken most of the discussion of ibn Yunis from King's paper.

father Abd al-Rahman was a distinguished historian. In addition to being a great astronomer, ibn Yunis was widely acclaimed as a poet, and some of his poems have survived. We do not know the date of his birth, but we do know that his earliest surviving observation is dated 977, and we know that he died in Cairo in 1009. He probably lived in Cairo all his life.†

As I just said, ibn Yunis's first known observation was made in 977, and he made observations fairly steadily from then to the solar eclipse of 1004 January 24. After 996, which saw the accession of Caliph al-Hakim, his work was sponsored by the caliph; and the tables which he prepared on the basis of his observations are dedicated to al-Hakim and in consequence are known as the Hakemite tables. These tables were finished in 1008 [ibn Yunis, 1008].

ibn Yunis explains that the Hakemite tables were meant to replace tables prepared by Yahya ibn Abi Mansur about 810. In the first part of his work, ibn Yunis reports a large number of observations, made both by himself and by earlier astronomers, which he compares with computations made from Yahya's tables. This comparison demonstrates the need for new tables, and also preserves much information about Islamic astronomy that would otherwise have been lost. ibn Yunis shows detailed knowledge of the work of many earlier astronomers, but he and al-Biruni show no knowledge of each other.

In particular, ibn Yunis was well acquainted with the work of Hipparchus and Ptolemy. He did not realize that Ptolemy's observations were fabricated but he did conclude that Ptolemy was an inferior observer to Hipparchus. As a result, he discarded Ptolemy's observations. He found that all other observations fitted adequately to a model in which the equator and equinox behave smoothly, and thus he did not have to work with models that included "trepidations" of the equator and equinox.

We know from the table of contents that ibn Yunis's zij, that is, the Hakemite tables, originally contained 81 chapters, but only about 50 still survive. The work that is cited in the references under ibn Yunis contains the text of the introduction, the table of contents, and the reports of the observations. This compilation was prepared in 1804 by Caussin de Perceval, and was published by him along with a parallel translation into French.‡ I have analyzed all the observations in Caussin's compilation that pertain to the sun and moon [Newton, 1970] and to Mercury, Venus, and Mars

---

†King says that he died in Fustat, but this is just the old half of what is now called Cairo. The new half, which was called Cairo from its founding, was established about 969. I do not distinguish between the two halves.

‡King [1976] says that this work was done by Armand-Pierre Caussin de Perceval. Actually, it was done by J. J. A. Caussin de Perceval, the father of Armand-Pierre.

[Newton, 1976]. Here, I will analyze again the observations that pertain to the moon, for two reasons. First, I will use the DE102 ephemeris in the analysis in order to have consistency with the rest of this volume. Second, I will test some corrections to the text that are suggested by Knobel [1879]. In spite of its age, I was not aware of Knobel's work when I did the earlier analysis.

Hartner [1970] is my source for the personal life of al-Battani, who is described as "one of the greatest Islamic astronomers". His names are listed as Abu Abd Allah Muhammad ibn Jabir ibn Sinan al-Raqqi al-Harrani al-Sabi al-Battani, and he is known in medieval Latin literature as Albatenius, Albategni, or Albategnius. In addition to two patronymics, he is associated with four descriptive names. It is believed that he was born in or near Harran or Haran in the Mesopotamian region of present-day Turkey; if so, this would account for one of his names. He spent most of his life and made most of his observations in Raqqa, in present-day Syria; this accounts for another name. al-Sabi refers not to a place but to a religious cult now known as the pseudo-Sabians of Haran. The pseudo-Sabians were a pagan cult of star-worshippers who pretended to be Sabians in order to gain certain advantages under Islamic law; the true Sabians were a Christian or semi-Christian group. Since al-Battani is known to have been a Muslim, the name al-Sabi indicates that some of his ancestors must have been pseudo-Sabians.

al-Battani, which is the name that the astronomer came to be known by, has no known origin, and all attempts to derive it from a place name have failed. It has been suggested that it refers to a minor division of Haran, or perhaps even to a street there.

It is believed that al-Battani's father was a famous instrument maker named Jabir ibn Sinan al-Harrani. If so, this might explain al-Battani's interest in astronomy and his skill at devising new types of astronomical instruments. We know almost nothing of his life except what we can glean from his astronomical observations. His first observation was made in 877, so it is fairly safe to say that he was born about 857. His last known observation was made in 918, when he was about 60, but there is an old record which says that he lived 11 years longer. According to this record, he was a member of a delegation from Raqqa to Baghdad, and he died in 929 while en route back to Raqqa. In contrast to ibn Yunis, al-Battani accepted and used Ptolemy's observations.

Our final source is al-Khwarizmi, for whom I rely mostly upon Toomer [1973]. His full name is Abu Ja'far Muhammad ibn Musa al-Khwarizmi. His name indicates that he was from Khwarizm, the birth region of al-Biruni, but some early sources call him al-Qutrubbulli, referring to a region near Baghdad. In order to reconcile these names, Toomer suggests that al-Khwarizmi came from Qutrubbull and that Khwarizm was the home of his ancestors.

All we know of his birth is that it was before 800. All that we know of his death depends upon a story that may not be reliable. According to this story, al-Khwarizmi was one of a group of astronomers (read astrologers in modern English) who were summoned to the sickbed of Caliph al-Wathiq in 847. As Toomer words it, they " . . predicted on the basis of the caliph's horoscope that he would live another fifty years and were confounded by his dying in ten days." If this story is true, al-Khwarizmi survived the caliph.

al-Khwarizmi wrote a manual of practical mathematics that became widely known in medieval Islam and Europe by one of the words, namely al-jabr, in its title.† His name also gave rise to the modern mathematical terms algorism and algorithm.

al-Khwarizmi's work on astronomy was called Zij al-sindhind. A zij, as we have seen, is a set of astronomical tables, sindhind is a corruption of a Sanskrit word, and the Zij al-sindhind was prepared by taking some elements from Hindu astronomy and some from Ptolemy. This zij was translated into Latin by (presumably) Adelard of Bath early in the twelfth century. Unfortunately for the history of astronomy, this translation is based not upon the original but upon a revision made by Islamic astronomers in Spain in the eleventh century. Nonetheless, al-Khwarizmi's zij is the earliest Muslim astronomical work that has survived to a considerable extent and in at least some resemblance to its original form.

The form of the tables is close to that used by Ptolemy, but the underlying models are those of Hindu astronomy, which was much less advanced than the ancient Greek astronomy. Toomer writes: "Nowhere in the work is there any trace of original observation or of more than trivial computation by the author." Most of the parameters are also Hindu in origin, with the notable exception of the obliquity. As Toomer says, this is strange because al-Khwarizmi is known to have discussed some Islamic observations made to determine the obliquity which led to the rather accurate value 23° 33'. Yet al-Khwarizmi adopted the far less accurate Ptolemaic value 23° 51'.

I gather that al-Khwarizmi is not of the stature of the other Islamic astronomers who have been discussed. His fame seems to come from his having been one of the earliest Muslims who attempted to master the existing systems of astronomy, systems that were described in languages foreign to him. Since he had so little writing in his own language to help him, perhaps we should praise his attempts and not play

---

†Al-jabr in this context refers to the process of eliminating negative quantities in an algebraic relation. The example given is to transform $x^2 = 40x - 4x^2$ into $5x^2 = 40x$. It should be emphasized that al-Khwarizmi did not know how to use these symbols, and that he had to express these relations in words.

him down for his failure to make original contributions. In
a parallel situation, medieval Europeans had to study
earlier astronomy for a long time before they attained the
state of being original.

2. ibn Yunis's Solar and Lunar Tables

Before taking up the actual observations made by Islamic astronomers, I will take up in this section the mean
longitudes of the sun and moon that are given in the
Hakemite tables of ibn Yunis [1008]. I will take up the
actual observations in later sections.

On page 222 of the cited edition and translation, ibn
Yunis gives the mean longitudes of the sun and moon for the
epoch of mean noon, Cairo time, on the first day of the Muslim year 391. The date is the equivalent of 1000 November
30,[†] and Cairo is 31°.25 east of Greenwich. Hence ibn
Yunis's epoch has the Julian day number 2 086 641.913 2,
Greenwich mean time.[‡] At this epoch, ibn Yunis's values
for $\underline{L}_S$ and $\underline{L}_M$, the mean longitudes of the sun and moon, are:

$$\underline{L}_S = 254;45,57,06 = 254°.765\ 86,$$

$$\underline{L}_M = 270;41,12,25 = 270°.686\ 78.$$

(V.2)

I want to make two principal changes from my earlier
treatment of those values [Newton, 1970, pp.31, ff], aside
from using Van Flandern's new mean elements [Van Flandern,
in preparation]. First, I took the reference epoch for the
accelerations $\underline{y}$ and $\underline{\dot{n}}_M$ to be 1900. Here, as I have already
explained, I take the epoch to be 1969 for $\underline{\dot{n}}_M$ and 1790 for
$\underline{y}$. Second, the tabular quantities I used were $\underline{L}_S$ and $\underline{L}_M$,
as given in Equations V.2. Here, I want to use the lunar
elongation $\underline{D}$ in place of $\underline{L}_M$. My reason for this comes from
a greater understanding of the procedures used by ancient
and medieval astronomers.

A common procedure for finding the orbital parameters
of the sun was to measure the times of three of the cardinal
points of the year, typically the two equinoxes and the
summer solstice. The analysis of these measurements gives

---

†Actually, at least in the medieval period, there is an
 ambiguity of a day in relating the Muslim calendar to the
 Julian calendar because of three different conventions
 about what day was the first day of the Muslim era. See
 Newton [1972a]. We can resolve this ambiguity only if
 there is additional information besides the calendar date,
 such as the day of the week; the Muslim week is the same as
 ours. In this case, only the date 1000 November 30 yields
 agreement with the astronomical data.

‡However, there is another complication in assigning this
 epoch that I will take up in a moment.

the eccentricity and apogee[†] of the sun, as well as the mean
longitude at some epoch. In this analysis, it is necessary
to have only a preliminary estimate of the mean motion. An
accurate value of the mean motion is then found by combining
the mean longitude with that found by some earlier astronomer. ibn Yunis found his mean motion by combining his
observations with those of Hipparchus.

The mean longitude L_M of the moon was found by measuring the times of the centers of lunar eclipses, when the
elongation of the moon is 180°. This means that the independent quantity in the lunar tables is the elongation D
rather than the mean longitude itself. We can see this by
taking the work of Ptolemy as an example. His value for the
mean motion n_S of the sun, which he adopted from Hipparchus,
is 36 000.329 degrees per Julian century, and his value for
the mean motion n_M of the moon is 481 267.360 degrees per
century. These values are based upon observations with an
average date somewhere near −200.

When we calculate the mean motions from modern theory,
using the accelerations found in volume 1, we get n_S =
36 000.741 and n_M = 481 267.762. Ptolemy's values are both
too small by about 0°.4 per century. However, his value for
the mean motion in elongation, which was taken from Babylonian astronomy, is 445 267.032 degrees per century, while
the value from modern theory is 445 267.021. The error in
Ptolemy's value is about 0°.01, about 1/40 of the error in
n_M. The mean elongation is clearly the independent value,
and the error in n_M is dominated by the large error in n_S,
which was added to a small independent error in the elongation.

Thus, instead of using the value of L_M in the second of
Equations V.2, I use the corresponding value of D. This is
clearly 15°.920 92. Before we use this value and the value
of L_S, we must correct for aberration. ibn Yunis did not
know about aberration, and so his coordinates are apparent
rather than geometric. To get the geometric values, we must
add the solar aberration at mean distance, namely 0°.005 69,
to L_S and we must subtract the same amount from D; we may
neglect the lunar aberration. Thus the values of L_S and D
to compare with modern results are

$$L_S = 254°.771\ 55,$$
$$D = \quad 15°.915\ 23.$$

(V.3)

[†]Ancient and medieval astronomers commonly used apogee
instead of perigee.

Equation I.24 gives Van Flandern's expression for $\underline{L}_S$ and Equation I.26 gives his expression for $\underline{D}$, after we change $\dot{n}_M$ from -26.21 to -28. When we use Equation I.24, we find that 2 086 641.952 9 is the dynamical time when $\underline{L}_S$ equals 254°.771 55. When we use Equation I.26, we find that 2 086 641.935 1 is the dynamical time when $\underline{D}$ equals 15°.915 23. The latter time, of course, depends upon $\dot{n}_M$, and the reader should have no difficulty in deriving the value of the time for any other value of $\dot{n}_M$.

That is to say, we get ΔT = 3430 seconds from the value of $\underline{L}_S$ and we get ΔT = 1892 seconds from the value of $\underline{D}$. Before we use these numbers, we must attach standard deviations to them. The value of $\underline{L}_S$ was found from the times of several cardinal points of the year, each with an uncertainty of perhaps an hour, so I think it is reasonable to take 2000 seconds as the standard deviation of the value of ΔT found from $\underline{L}_S$. The uncertainty in the value found from $\underline{D}$ is probably dominated by another consideration, namely our ignorance of the epoch that ibn Yunis meant by noon, Cairo mean time.

In my earlier use of these mean coordinates, I took it for granted that ibn Yunis used the same definition of mean time that we do, but I realize now that this may not be so. To put the problem explicitly, we get mean time from a fictitious entity called the mean sun, which moves uniformly in the equatorial plane and whose right ascension equals that of the real sun on the average. If we use α_F to denote the right ascension of the fictitious mean sun and α_S to denote the right ascension of the real sun, we define the equation of time $\underline{E}$ as

$$\underline{E} = \alpha_F - \alpha_S, \tag{V.4}$$

after conversion into time units. Thus the average value of $\underline{E}$ is zero.

Earlier astronomers did not necessarily do this, and there are many definitions of the equation of time. Since mean time is derived from the time kept by the real sun (apparent time) by using the equation of time, a difference in the equation of time means a difference in the value of mean time that an astronomer would assign to a particular event. The various definitions of $\underline{E}$ differ from Equation V.4 by an arbitrary constant. That is, to generalize $\underline{E}$, we take it to be

$$\underline{E} = \alpha_F - \alpha_S + \underline{C}. \tag{V.5}$$

The choice of $\underline{C}$ was up to the individual astronomer before the days of the International Astronomical Union, which has chosen $\underline{C}$ = 0. ben Gerson, as we saw in Section IV.2, chose the value of $\underline{C}$ that makes $\underline{E}$ have a minimum value of 0. Ptolemy, in effect, chose $\underline{C}$ so that $\underline{E}$ is zero at an arbitrary epoch.† We cannot assume that ibn Yunis chose C = 0, and we would like to know his choice.

Unfortunately, we may never know. ibn Yunis discusses the equation of time in Chapter 3 of his work, as we can see from the published table of contents. However, Caussin started the cited translation of ibn Yunis with Chapter 4. About the same time as Caussin [Kennedy, 1956, p. 126], one J. J. Sédillot made a more extensive translation of ibn Yunis, including his Chapter 3. This translation was never published and has in fact disappeared, but Delambre [1819, pp. 76 - 156] made an extensive astronomical analysis of its contents. According to Delambre's discussion (page 98), ibn Yunis describes five different definitions of the equation of time, but apparently without ever telling us the definition that he used.‡ I hope somebody with the necessary access and linguistic ability will study a manuscript of Chapter 3 and tell us if it gives ibn Yunis's value of C.

Even without an explicit statement by ibn Yunis of his value of C, we could still determine it from a single instance in which he gives both the apparent and mean times of a specific event, as he might well do in the course of reducing an observation. I have searched through the published but partial translation of his work, and also the analysis by Delambre, without finding a single instance. Thus, at least for the time being, we do not know what ibn Yunis meant by noon, Cairo mean time.

It is unlikely that he chose a value of C greater in magnitude than the largest absolute value of $\overline{\alpha}_F - \overline{\alpha}_S$, which is about $16^m 20^S$ in time units, but he may well have chosen a value this large. If we take C = 0 in the analysis, which is the best choice we can make, we may thus make an error as large as $16^m 20^S$ (= 980 seconds) in the value of ΔT. The contribution of this error to the standard deviation of ΔT should be less than this, but there are other sources of error that I have not discussed. All things considered, I think it is reasonable to take 900 seconds as the standard deviation of the value of ΔT derived from the lunar elongation D.

When we take the values of ΔT stated a moment ago (which tacitly assume C = 0), combined with the standard deviations we have chosen, and use Equation I.15, we get

---

†Actually, Ptolemy does not deal directly with E itself, but only with intervals measured in mean time and in apparent time. This makes his treatment of the equation of time [Ptolemy, ca. 142, Chapter III.9] rather confusing. See Pedersen [1974, pp. 154ff].

‡In fact, Delambre writes: "Ebn Jounis tells us less (about the equation of time) than Theon." The parenthesis is mine. Theon is a minor Alexandrian astronomer of the fourth century. See Section VI.1.

$$\underline{y} = -34.9 \pm 18.3 \qquad \text{from } \underline{L}_{\underline{S}},$$
$$\underline{y} = -19.3 \pm 9.2 \qquad \text{from } \underline{D}. \tag{V.6}$$

The second of these depends upon the value of $\dot{n}_M$ but the first does not.

3. Islamic Records of Solar Eclipses

These records have been used at least twice before in studies of the secular accelerations [Newcomb, 1875 and Newton, 1970]. I study them again here for two reasons. One reason is to make their analysis consistent with that of the other records by using the DE102 ephemeris. The other is to test some suggestions about the texts that were made by Knobel [1879].

On the basis of his experience in working with Arabic manuscripts, Knobel concluded that certain numerals in Arabic are more likely than others to be changed accidentally in reading. These changes may happen either in copying a manuscript or in translating one. In his paper, Knobel discusses some of the difficulties that Newcomb found in trying to use the records of ibn Yunis [1008], and he suggests alternate readings in many places to those adopted by Newcomb. I will see whether we can settle among the suggested readings on the basis of calculation.

The dates of all the records are given in either the Muslim calendar or the Persian one. In most of the cases, the year, month, and day are given, but in some cases only the year and month are given. In addition, the day of the week is often given. I believe that there is only one source of ambiguity in translating the dates into the Julian calendar. This is the fact, mentioned in the preceding section, that there are three different conventions about the first day of the Muslim era, but the three conventions yield consecutive days. Thus, if we use the middle one of the three possibilities, the maximum uncertainty is one day. Since we are dealing with solar eclipses visible at known points, we can identify the eclipses without question. Hence, I give only the dates in the Julian calendar, without going into the process by which the dates are determined.

829 November 30. Source: ibn Yunis [1008, p. 84]. This observation was probably made by an early Islamic astronomer named Habash, who prepared a well-known set of tables. It was observed at Baghdad, where the altitude of the sun was 7° at the beginning of the eclipse and 24° at the end.

866 June 16. Source: ibn Yunis [1008, p. 92]. This observation was made in Baghdad by an astronomer named Mahani. The eclipse should have begun at 6;3 unequal hours, but it was late by more than a third of an hour. The middle at 7;26 and the end at 8;30. The eclipsed part was between 7 and 8 digits of the diameter.

Here we have the same kind of hour that we encountered in Section IV.6. An hour of the night means a twelfth of

the interval between sunset and sunrise while an hour of the
day means a twelfth of the interval between sunrise and
sunset. In the remaining records of this work, I will use
"unequal hour" to mean either kind; the context will make it
clear whether the hour is of the day or of the night. I
will use "equal hour" to mean 1/24 of a day, and if I do not
use a modifier the equal hour is to be understood.

Newcomb [1875, p. 46] omits this record and goes directly from the solar eclipse of 829 November 30 to that of
923 November 11.† I do not know whether this is an oversight or whether he omitted it because he thought that the
data were predicted ones and not observed ones. I agree
that the wording makes it unclear whether the middle and end
times are observed or not, but I think the record makes it
clear that at least two numbers in it are observed. One is
the beginning time, which is more than a third of an hour
after 6;3 hours. I think we may safely take "more than a
third" to be also "less than a half", and thus we may take
the time to be 25 minutes after 6;3 hours, or 6;28 hours,
unequal time. The other measured quantity is the magnitude,
which we take as 7 $\frac{1}{2}$ digits, that is, 7 $\frac{1}{2}$ twelfths, of the
diameter.

Now let us look at the times stated for the middle and
end of the eclipse. If these times are calculated, the middle should agree closely with the average of 6;3 and 8;30,
but it differs by 10 minutes. However, it agrees closely
with the average of 6;28 and 8;30, so I feel we are safe in
taking all the numbers quoted to be observed ones. In my
earlier analysis [Newton, 1970, pp. 146, 247], I used the
middle and end times but not the beginning. I did not record the reasoning back of this decision and I cannot now
reconstruct it. Perhaps the omission of the beginning was
an oversight.

I did not use the magnitude of this eclipse [pp. 241ff]
because it poses a peculiarity that I could not cope with
using my earlier crude method of analysis. Here its use
should pose no computational problem.

891 August 8. Source: al-Battani [ca. 925, Chapter
XXX]. al-Battani observed this eclipse in Raqqa. The magnitude was greater than 8 digits according to sight and the
middle of the eclipse came at 1 unequal hour after noon.
al-Battani does not tell us how he measured the time nor how
he judged the instant when the middle of the eclipse came,
and we can only hope that he used sound procedures.

Sometimes al-Battani states a magnitude to be a certain
number of digits of the diameter. Here he says in contrast
that the magnitude is measured "according to sight" (secun-

---

†That is, he goes directly so far as solar eclipses are
concerned. He lists all recorded eclipses chronologically
and thus he has some lunar eclipses in the interval. Newcomb uses the old astronomical day that begins at noon, and
the dates he assigns sometimes differ from mine by a day.

dum visum in the cited Latin translation). I think we are
fairly safe in taking this to be the magnitude of the area.
See Appendix II for the translation of this into the magnitude of the diameter. I take "greater than 8 digits" to be
8 ½ digits of the area, or 0.76 of the diameter.

901 January 23a. Source: al-Battani [ca. 925, Chapter
XXX]. We have two records of this eclipse, which I distinguish by putting a and b after the date. This observation
was made by an unnamed astronomer in Antakya (Antioch), Turkey. There, the magnitude was greater than 6 digits, and
the middle of the eclipse was 3 2/3 equal hours before noon.
I take the magnitude to be 6 ½ digits of the area, which
equals 0.63 of the diameter.

901 January 23b. Source: al-Battani [ca. 925, Chapter
XXX]. This observation was made by al-Battani himself in
Raqqa. The magnitude was less than 8 digits, which I take
to be 7 ½ digits of the area, or 0.70 of the diameter. The
middle of the eclipse was 3 ½ equal hours before noon.

923 November 11. Source: ibn Yunis [1008, p. 114].
This observation was made in Baghdad by two astronomers,
father and son, named Amajour. The height of the sun was 8°
at the middle of the eclipse and 20° at the end; the sun was
to the east. It is also stated that the end was at $2^h 12^m$
in unequal hours, presumably after sunrise. The magnitude
was 3/4 of the diameter.

928 August 18. Source: ibn Yunis [1008, p. 120].
This observation was also made by both Amajours in Baghdad.
They say that they watched this eclipse by reflection from
water, although they had used pinholes to observe others.
By calculation, the sun should have risen eclipsed, and it
did rise eclipsed by a little less than a fourth of its surface. I take this magnitude to be 2 3/4 digits of the area,
or 0.34 of the diameter. At the end of the eclipse, the
height of the sun was 12° to the east, less a third of a
division, which was itself a third of a degree. That is,
the height was 11 8/9 degrees.

Except for the record of 1019 April 8, all the remaining Islamic records of solar eclipses are from ibn
Yunis, and I will cite only the page number. Further, all
these observations were made by ibn Yunis in Cairo, sometimes with many others observing simultaneously. Before I
present his records, I want to describe a usage he has that
I do not remember seeing elsewhere. Sometimes he refers to
the beginning of an eclipse, sometimes he refers to the time
when an eclipse began to appear, and sometimes he refers to
both times for the same eclipse. When this happens, the
beginning is earlier than the time when the eclipse began to
appear. I think this usage means the following: ibn Yunis
realized that he could not tell that an eclipse had begun
until the darkness had made a finite intrusion into the
bright disk, but this did not happen until some time after
the two disks had touched each other. By the beginning he
meant the instant of contact, and he tried to estimate the
instant of contact from the time when he first saw the
eclipse.

ibn Yunis also refers to a phase that the translator Caussin renders as attouchement. In my earlier study, I did not want to get into the question of what this phase might be, and I ignored all times of attouchement. Here I will study them to see if we can determine the meaning. If the French has preserved accurately the sense of the original, attouchement should mean the instant of contact. That is, it should be the beginning, and in one or two places ibn Yunis is quoted as saying "attouchement, je veux dire le commencement", or some such phrase. If this meaning seems to be consistent with the data, I will use the times of attouchement in this study.

I have not had an opportunity to experiment with eclipses to determine the magnitude when an eclipse begins to appear. In an earlier work [Newton, 1970, p. 125] I did some simulated experiments and found that I could always see an eclipse when the magnitude was 0.01 and that I could never see it when the magnitude was 0.005. Hence, I took 0.0075 as the magnitude when an eclipse was first seen. On further reflection, I believe that I should use 0.01, since an observer would always need a little time to find the eclipsed region after it became big enough to be seen. All these numbers refer to the magnitude of the diameter.

ibn Yunis sometimes refers to the end of an eclipse and sometimes to the time when it could no longer be seen, or some such term. He takes these to be the same thing, and I believe this is reasonable. Let us suppose that my experiments just described are valid, and that one cannot see an eclipse any longer when the magnitude has dropped to 0.005. However, I think an observer would tend to wait a little longer to be sure that the eclipse could no longer be seen, and this would take him close to the end of the geometric contact.

When ibn Yunis gives both the time of the beginning and the time when an eclipse began to appear, the latter time is clearly the measured one. This poses a slight problem: At the latter time, we have a definite time but an uncertain magnitude, while at the beginning we have a definite magnitude but an uncertain time. When ibn Yunis gives it, I will use the time when the eclipse began to appear, taking the magnitude then to be 0.01 of the diameter. When ibn Yunis gives only the beginning, we have no way to know how he determined it, and we will have to take it as given.

977 December 13. Page 164. The eclipse began to appear when the height of the sun was between 15° and 16°; I will use 15 $\frac{1}{2}$ degrees. At the end of the eclipse, the height was 33° 20', "everyone being in agreement about the end." That is, several people observed this eclipse in company. The magnitude was about 8 digits of the diameter, less than 7 of the area. From Figure A.II.3 in Appendix II, we see that 8 digits of the diameter equals almost exactly 7 digits of the area. Since the record says definitely less than 7 digits of the area, and is slightly vague about 8 digits (0.67) of the diameter, I take the magnitude to be 0.65 of the diameter.

978 June 8. Page 166. The height of the sun when the eclipse began to appear was about 56°, and the height at the end was about 26°. The magnitude was 5 ½ digits of the diameter, or 4;10 digits of the area. This is ibn Yunis's conversion, and it agrees accurately with Ptolemy's table [Ptolemy, ca. 142, Chapter VI.8] and with Figure A.II.3.

979 May 28. Page 168. The height of the sun when the eclipse became sensible to the sight was 6 ½ degrees. The magnitude was 5 ½ digits of the diameter. The sun set eclipsed. I think that "becoming sensible to the sight" is the same thing as "appearing to begin". The record does not tell us whether 5 ½ digits was the maximum magnitude or the magnitude at sunset, but we should be able to settle this by computation.

985 July 20. Page 172. The height of the sun at the beginning of the eclipse was about 23°, and the height at the end, when the eclipse was no longer sensible to the sight, was about 6°. The magnitude was one-fourth of the diameter. Here "the end" and "no longer sensible to the sight" are explicitly equated.

993 August 20. Page 174. The height of the sun was 27° to the east at the beginning, 45° at the maximum, and 60° at the end. The magnitude was 2/3 of the area. By Figure A.II.3, this is about 0.73 of the diameter.

Both here and in the record of 860 June 16, we have a time given for the middle or for the time of greatest phase, and this time seems to be an independent measurement; it is not found by averaging the beginning and end. I do not see how the middle times can be judged as accurately as the beginning or end times, and I will treat the middle times with caution in the analysis.

1004 January 24. Page 178. This is the last solar eclipse observed by ibn Yunis. The magnitude was 11 digits. The height of the sun was 16° 30' in the west when the eclipse began to appear, 18° 30' at the beginning, 15° when a fourth of the diameter was eclipsed, 10° when half the diameter was eclipsed, and 5° at the time of greatest eclipse. The sun presumably set eclipsed, but ibn Yunis does not give the magnitude at this time.

ibn Yunis does not say whether the 11 digits were of the diameter or the area. However, the other measurements of the magnitude explicitly refer to the diameter, and I assume that the maximum magnitude does also.

Here we see both the appearance of the eclipse and its beginning, and the beginning is earlier. Knobel [1879, p. 340] says that we should read 17° 30' instead of 18° 30' for the beginning, which still leaves the beginning earlier than the first appearance. Since I will not use this time, the correction has no consequence for our purposes, but I will test Knobel's suggested correction.

1019 April 8. Source: al-Biruni [1025, p. 261]. "On the morning of this eclipse", al-Biruni writes, " . . we

were near Lamghan, between Qandahar and Kabul, in a valley
surrounded by mountains, where the sun could not be seen
unless it was at an appreciable altitude above the horizon.
At sunrise, we saw that approximately one-third of the sun
was eclipsed and that the eclipse was waning."

TABLE V.1

ISLAMIC RECORDS OF SOLAR ECLIPSES

Date	Place	Magnitude[a]	Phase[b]	Time[c]
829 Nov 30	Baghdad		$\underline{B}$ $\underline{E}$	$a = 7°$ $\underline{a} = 24°$
866 Jun 16	Baghdad	0.62	$\underline{B}$ $\underline{M}$ $\underline{E}$	6;28 unequal hrs 7;26 unequal hrs 8;30 unequal hrs
891 Aug 8	Raqqa	0.76	$\underline{M}$	1 unequal hr after noon
901 Jan 23a	Antakya	0.63	$\underline{M}$	3 2/3 equal hrs before noon
901 Jan 23b	Raqqa	0.70	$\underline{M}$	3 1/2 equal hrs before noon
923 Nov 11	Baghdad	0.75	$\underline{M}$ $\underline{E}$ $\underline{E}$	$a = 8°$ $\underline{a} = 20°$ $\overline{2};12$ unequal hrs
928 Aug 18	Baghdad		m=0.34 $\underline{E}$	sunrise $\underline{a}$=11 8/9 degrees

[a]The magnitude of the diameter.

[b]In this column, B means the beginning, M the middle or
maximum, and E the end. A means the phase that ibn Yunis
calls appearance. m denotes an instantaneous value of the
magnitude of the diameter.

[c]In this column, a denotes the altitude of the sun.

TABLE V.1 (continued)

Date	Place	Magnitude[a]	Phase[b]	Time[c]
977 Dec 13	Cairo	0.65	A	a=15 1/2 degrees
			E	a=33° 20'
978 Jun 8	Cairo	0.46	A	a=56°
			E	a=26°
979 May 28	Cairo	0.46[d]	A	a=6 1/2 degrees
985 Jul 20	Cairo	0.25	B	a=23°
			E	a= 6°
993 Aug 20	Cairo	0.73	B	a=27°
			M	a=45°
			E	a=60°
1004 Jan 24	Cairo	0.92	B	a=18° 30'[e]
				a=17° 30'[e]
			A	a=16° 30'
			m=0.25	a=15°
			m=0.50	a=10°
			M	a=5°
1019 Apr 8	Kabul		m=0.39	40 minutes after sunrise

[a]The magnitude of the diameter.

[b]In this column, B means the beginning, M the middle or maximum, and E the end. A means the phase that ibn Yunis calls appearance. m denotes an instantaneous value of the magnitude of the diameter.

[c]In this column, a denotes the altitude of the sun.

[d]We cannot tell from the record whether this is the greatest magnitude or the magnitude at sunset.

[e]Alternative readings of the text.

I cannot locate any of the places mentioned except Kabul, so I use the coordinates of Kabul in the analysis. We do not know whether the magnitude is of the diameter or the area. When I used this record before [Newton, 1972a], I took the magnitude to be of the diameter. I believe that it is better to take the average of the two possibilities. One possibility is that the magnitude means one-third of the diameter. The other is that it means one-third of the area, which by Figure A.II.3 equals 0.45 of the diameter. I use the average of the two possibilities, namely 0.39 of the diameter.

Earlier, I arbitrarily assumed that the sunrise was delayed by 40 ± 20 minutes, and I see no way to make a less arbitrary choice. The uncertainty in the time probably outweighs the other uncertainties in the record.

Table V.1 summarizes the data found in the preceding records. The first two columns of the table give the date of an eclipse and the place of its observation. The third column gives the maximum magnitude of the eclipse for these records that give an estimate of it. I have converted to the magnitude of the diameter for those records that give the magnitude of the area, and I have converted from digits or a fractional form to decimal form.

The last column gives the measured time of the phase of the eclipse that is given in the fourth column. More precisely, the last column gives the time when the time is directly stated, and otherwise it gives the basic data from which we may infer the time.

4.   Islamic Records of Lunar Eclipses

The remarks that were made at the beginning of the preceding section also apply here, with the obvious change of "solar" to "lunar".

854 February 16. Source: ibn Yunis [1008, p. 86]. The eclipse was observed by Mahani in Baghdad, and it began $10^h 3^m$ after noon. The kind of hour is not stated. However, 10 hours after noon is necessarily after sunset, and if the hours were unequal, they would have to be a peculiar mixture of hours of the day and hours of the night. Hence I believe that the time is in equal hours, but that it is in apparent time rather than mean time. The uneclipsed part of the disk exceeded one-tenth. This means, I believe, that the magnitude of the area was less than 0.9. From Figure A.II.2, we see that a magnitude of 0.9 of the area for a lunar eclipse equals 0.86 of the diameter. I assume as before [Newton, 1970, p. 148] that the magnitude was 0.83 of the diameter. This probably means that the magnitude was originally recorded as 10 digits of the diameter, and that someone converted this precise statement into a vague statement about the magnitude of the area.

Newcomb [1875] did not mention this record.

854 August 12. Source: ibn Yunis [p. 88]. This was also observed by Mahani in Baghdad. At the beginning of the eclipse the height of Aldebaran (α Tauri) was 45° 30′ in the east. The record emphasizes that this was the only circumstance of the eclipse that was measured, but that this measurement is highly accurate.

856 June 22. Source: ibn Yunis [p. 90]. This was also observed by Mahani in Baghdad. At the beginning of the eclipse the height of α Tauri was 9° 30′ in the east. The uneclipsed part was between 1/3 and 1/4 of the diameter.

That is, the magnitude was between 2/3 and 3/4. This probably means that the estimated magnitude was 8 $\frac{1}{2}$ digits, or 0.71. We do not know whether this is meant to apply to the area or to the diameter, but luckily it does not matter. The magnitudes of the area and of the diameter (Figure A.II.2) are almost exactly equal at a value of 0.71, so we can safely take the magnitude of the diameter to be 0.71.

866 November 26. Source: ibn Yunis [pp. 92ff]. ibn Yunis quotes Mahani as saying that the calculated time of opposition at Baghdad was 9^h 31^m, in unequal hours at Baghdad, that the sun was at 8° 31' in Sagittarius, that the node was at 19° 50' of Gemini, and that the latitude of the moon was 59' south. The beginning, he says, was at 8^h 55^m and the end was at 10^h 7^m 30^S, both in unequal hours. The magnitude was 1 $\frac{1}{2}$ digits of the diameter. The calculated time of opposition is the average of the times given for the beginning and end, and it is likely that all the times in this quotation are calculated. It is possible that the magnitude is an observed quantity, but I think it is safer to treat it as calculated also. I calculated the circumstances of this eclipse in the earlier analysis, while giving them negligible weight. Here I ignore them entirely, and Newcomb does the same.

883 July 23. Source: al-Battani [ca. 925, Chapter XXX]. al-Battani observed this eclipse in Raqqa. The middle of the eclipse was at slightly more than 8 equal hours after noon; I take this to be 8^h 10^m after noon, apparent time at Raqqa. The magnitude was more than 10 digits of the diameter; I take this to be 0.88.

901 August 3a. Source: al-Battani [ca. 925, Chapter XXX]. al-Battani gives two records of this lunar eclipse, just as he did with the solar eclipse of 901 January 23, and I distinguish the records by letters a and b following the dates. This observation was made in Antakya, where "a very little less than its diameter was eclipsed." The middle of the eclipse was at 15 1/3 equal hours after noon. I take the magnitude to be 11 $\frac{1}{2}$ digits; I used 11.6 digits in the earlier analysis.

901 August 3b. Source: al-Battani [ca. 925, Chapter XXX]. al-Battani observed this eclipse in Raqqa. Here too, a very little less than the diameter was eclipsed. The middle of the eclipse was at 15^h 35^m, equal hours, after noon.

923 June 1. Source: ibn Yunis [p. 112]. This eclipse was observed by both Amajours in Baghdad. The moon rose eclipsed by 3 digits of the diameter or more; I take this to be 3 $\frac{1}{4}$ digits. The middle of the eclipse came at 1^h 40^m in equal hours of the night; I take this to mean 1.67 equal hours after sunset. The end of the eclipse was at 3 equal hours of the night, when the height of α Cygni was 29° 30'

to the east. The magnitude was greater than 9 digits of the diameter, which I take to be 9 $\frac{1}{4}$ digits.

925 April 11. All the remaining records of lunar eclipses except the last one are from ibn Yunis, and from now on I will cite only the page number; this record is on page 116. This was observed at Baghdad by one of the Amajours, but which one is not stated. The eclipse was total. At the beginning, the height of α Bootis was 11° to the east, and at the end the height of α Lyrae was 24°. The record goes on to say that the times, according to the observations, were 55 minutes, in unequal hours, at the beginning and 4^h 36^m, in unequal hours, at the end.

927 September 14. Page 118. This was observed by the younger Amajour in Baghdad. At the beginning, the height of α Canis Majoris was 31° to the east; from this, the observer takes the time to be 10 unequal hours after sunset. The magnitude was between 3 and 4 digits of the diameter, which I take to be 3 $\frac{1}{2}$ digits.

929 January 27. Page 122. This was observed in Baghdad, apparently by the father Amajour. The eclipse was total. At the beginning, the height of α Bootis was 18° to the east, which made the time 5 unequal hours after sunset. As Newcomb [1875, p. 54] says, the altitude of α Bootis cannot be correct. Knobel [1879, p. 339] says that the numbers 18, 33, and 38 are particularly likely to be confused in Arabic script, and that the recorded altitude must have been either 33° or 38°. I will do the calculations for all three readings to see if we can reach a conclusion.

933 November 5. Page 124. This was observed in Baghdad, apparently by the father Amajour again. The eclipse was total. At the beginning, the height of α Bootis was 15° to the east, so the time was 9^h 56^m, in unequal hours, after sunset.

979 May 14. Page 168. All the remaining observations in this section, except the last one, were made by ibn Yunis in Cairo. On 979 May 14, the moon rose eclipsed. The magnitude was between 8 and 9 digits of the diameter, which I take to be 8 $\frac{1}{2}$ digits. The record is not clear whether this is the greatest magnitude or the magnitude at moonrise, but we may be able to settle this by calculation. The end of the eclipse came at 1^h 12^m, equal hours, of the night. We have none of the data by which this time was measured, so we must accept it at face value.

979 November 6. Page 168. The magnitude was 10 digits of the area, so a particular astronomer is not necessarily consistent in his usage. By Figure A.II.2, this is 0.80 of the diameter. The height was 64° 30' to the east at the

beginning and 65° to the west at the end. The record does not state the body to which these elevation angles refer but, to anticipate later results, we get consistent times if we take them to refer to the moon.

980 May 3. Page 170. The eclipse was total. The height of the moon at the beginning of the eclipse was 47° 40', but the direction is not stated. The end of the eclipse was at 36^m, equal time, before sunrise. As Newcomb [1875, p. 48] says, the altitude of the moon is impossible. Knobel [1879, p. 339] says that the number 47 may look rather like the number 41 in Arabic script, so I will try both 47° 40' and 41° 40'. We must accept ibn Yunis's value for the end of the eclipse.

981 April 22. Page 170. The height of the moon at the beginning was about 21°, the magnitude was about 3 digits of the diameter, and the end of the eclipse was about a quarter of an hour before sunrise. It does not matter whether an interval this short is in equal or unequal hours.

981 October 16. Page 170. The magnitude was 5 digits of the diameter. At the time of attouchement the height of the moon was 24°. I believe this is the first instance of the phase translated as attouchement, whose meaning we do not yet know.

983 March 1. Page 172. The eclipse was total. The height of the moon when the eclipse became sensible to the sight was 66°, and the height at the end was 35° 50'. Knobel [1879, p. 339] says that we should read 62° for 66°, so I will try both values.

986 December 19. Page 172. The height of the moon at the beginning of the visible eclipse was 24° to the west. ibn Yunis evaluated the height at attouchement to be 50° 30'. The magnitude was 10 digits of the diameter. Knobel [p. 339ff] says that we should read 28 for 50 in the second altitude, which would make the height at attouchement be 28° 30'. I will try both readings, but Knobel's revision seems likely; it puts attouchement a moderate time before the eclipse could actually be seen. The moon set eclipsed, but the magnitude at setting is not recorded.

990 April 12. Page 174. ibn Yunis writes that "the height of the moon at the beginning, I mean at the moment of attouchement, was 38°." Here, ibn Yunis says explicitly that beginning and attouchement are the same thing. The magnitude was 7 ½ digits. The end came at the rising of the first degree of Aquarius.

Knobel does not comment on this record, but he has already told us in connection with the eclipse of 929 January 27 that 33 and 38 are particularly likely to be confused.

Hence I will try both 33° and 38° for the height of the moon
to see if a choice can be made. An unknown scholiast has
already suggested the change to 33° in the copy of ibn Yunis
that I used.

1001 September 5. Page 174. All that ibn Yunis says
of this eclipse is that the end came at 2 unequal hours
after sunset. Newcomb seems to have overlooked this record.

1002 March 1. Page 176. This was a total eclipse. At
the beginning, the height of α Bootis was 52° to the east
and that of α Aurigae was 14° to the west. The height of
α Bootis at the end was 35°. These data pose so many problems that I threw up my hands in my earlier study [Newton,
1970, p. 151] and did not attempt to use this record at all.
Newcomb [1875, p. 50], however, did decide to derive a time
for the beginning, and his time looks reasonable.

Knobel [1879, p. 340] has considerable to say about
this record. To start with, he writes, the printed Arabic
says that the altitude of al-Simak was 12° (not 52°) to the
east at the beginning, and that the altitude of al-Ramih was
35° at the end. Then he writes: "Though the name 'Al-
Simak' is applied to another star besides Arcturus,† viz. α
Virginis, I cannot find the name 'Al-Ramih' given to any
other star except α Bootis." Thus, he suggests, the record
says that the altitude of α Virginis was 12° at the beginning and that the altitude of α Bootis was 35° at the end.

With regard to the star in the west, Knobel says that
its Arabic name is al-Hadi. In the notes to his translation, Caussin says that he cannot find this name in the Arabic catalogues of stars, but that he has found a similar
name applied to α Aurigae. Knobel says that Caussin missed
its occurrence, and that it is applied in Arabic to α Tauri.
Altogether, then, he says, we have α Virginis at 12° in the
east and α Tauri at 14° west at the beginning; this is a far
cry from Caussin's translation. He agrees with Caussin that
we have α Bootis at an altitude of 35° at the end. However,
since α Bootis has been confused with α Virginis, I will try
α Virginis as an alternate to α Bootis for the end observation.

Thus we have many permutations to test, and I will try
them all.

---

†Arcturus is a name for α Bootis that goes back to ancient
   Greek.

TABLE V.2

ISLAMIC RECORDS OF LUNAR ECLIPSES

Date	Place	Magnitude[a]	Phase[b]	Time[c]
854 Feb 16	Baghdad	0.83	B	10.05 equal hrs after noon
854 Aug 12	Baghdad		B	a(α Tau) = 45° 30′east
856 Jun 22	Baghdad	0.71	B	a(α Tau) = 9° 30′east
883 Jul 23	Raqqa	0.88	M	8.17 equal hrs after noon
901 Aug 3a	Antakya	0.96	M	15.33 equal hrs after noon
901 Aug 3b	Raqqa	0.96	M	15.58 equal hrs after noon
923 Jun 1	Baghdad	0.78	m=0.27 M	moonrise 1.67 equal hrs after sunset
			E	3 equal hrs after sunset[d] a(α Cyg) = 29° 30′ east[d]

[a]The magnitude of the diameter.

[b]In this column, B means the beginning, M the middle or maximum, and E the end. A means the phase that ibn Yunis calls appearance, and T means the phase that Caussin translates as attouchement. m denotes an instantaneous value of the magnitude of the diameter.

[c]In this column, a notation such as a(α Tau) means the altitude of α Tauri.

[d]Separate statements about the time of the same event.

TABLE V.2 (<u>continued</u>)

Date	Place	Magnitude[a]	Phase[b]	Time[c]
925 Apr 11	Baghdad		<u>B</u>	a(α Boo)=$-11°$ east[d] 0.92 unequal hours[d]
			<u>E</u>	a(α Lyr) = $-24°$[d] 4.60 unequal hours[d]
927 Sep 14	Baghdad	0.29	<u>B</u>	a(α CMa) = $-31°$ east[d] 10 unequal hrs after sunset[d]
929 Jan 27	Baghdad		<u>B</u>	a(α Boo)=$-18°$ east[d,e] a(α Boo) = $-33°$ east[e] a(α Boo) = $-38°$ east[e] 5 unequal hrs after sunset[d]
933 Nov 5	Baghdad		<u>B</u>	a(α Boo) = $-15°$ east[d] 9.93 unequal hrs after sunset[d]

[a]The magnitude of the diameter.

[b]In this column, B means the beginning, <u>M</u> the middle or maximum, and <u>E</u> the end. <u>A</u> means the phase that ibn Yunis calls appearance, and <u>T</u> means the phase that Caussin translates as <u>attouchement</u>. <u>m</u> denotes an instantaneous value of the magnitude of the diameter.

[c]In this column, a notation such as <u>a</u>(α Tau) means the altitude of α Tauri.

[d]Separate statements about the time of the same event.

[e]Alternative readings of the text.

TABLE V.2 (continued)

Date	Place	Magnitude[a]	Phase[b]	Time[c]
979 May 14	Cairo	0.71[e]	m=0.71[e]	moonrise
			E	1.20 equal hrs after sunset
979 Nov 6	Cairo	0.80	B	a(moon) = 64° 30'east
			E	a(moon) = 65° west
980 May 3	Cairo		B	a(moon) = 47° 40'[e]
				a(moon) = 41° 40'[e]
			E	0.6 equal hrs before sunrise
981 Apr 22	Cairo	0.25	B / E	a(moon) = 21° 0.25 hrs before sunrise
981 Oct 16	Cairo	0.42	T	a(moon) = 24°
983 Mar 1	Cairo		A	a(moon) = 66°[e]
				a(moon) = 62°[e]
			E	a(moon) = 35°50'
986 Dec 19	Cairo	0.83	A	a(moon) = 24° west
			T	a(moon) = 50° 30'[e]
				a(moon) = 28° 30'[e]

[a]The magnitude of the diameter.

[b]In this column, B means the beginning, M the middle or maximum, and E the end. A means the phase that ibn Yunis calls appearance, and T means the phase that Caussian translates as attouchement. m denotes an instantaneous value of the magnitude of the diameter.

[c]In this column, a notation such as a(α Tau) means the altitude of α Tauri.

[d]Separate statements about the time of the same event.

[e]Alternative readings of the text.

TABLE V.2 (<u>continued</u>)

Date	Place	Magnitude[a]	Phase[b]	Time[c]
990 Apr 12	Cairo	0.63	<u>B</u>	a(moon) = $\overline{38}$°e
				a(moon) = $\overline{33}$°e
			<u>T</u>	a(moon) = $\overline{38}$°e
				a(moon) = $\overline{33}$°e
			<u>E</u>	rising, 1st degree Aqr
1001 Sep 5	Cairo		<u>E</u>	2 unequal hrs after sunset
1002 Mar 1	Cairo		<u>B</u>	$a(\alpha$ Boo) = $\overline{52}$° east[d,e]
				$a(\alpha$ Boo) = $\overline{12}$° east[e]
				$a(\alpha$ Vir) = $\overline{52}$° east[e]
				$a(\alpha$ Vir) = $\overline{12}$° east[e]
			<u>B</u>	$a(\alpha$ Aur)[e] = $\overline{14}$° west[d,e]
				$a(\alpha$ (Tau) = $\overline{14}$° west[e]
			<u>E</u>	$a(\alpha$ Boo) = $\overline{35}$°e
				$a(\alpha$ Vir) = $\overline{35}$°e
1019 Sep 16	Ghazni		<u>A</u>	$a(\alpha$ Au$\underline{r}$) = $\overline{65}$°.8[d]
				$a(\alpha$ CMa) = $\overline{17}$°[d]
				$a(\alpha$CMi) = $\overline{22}$°[d]
				$a(\alpha$ Tau) = $\overline{63}$°[d]
		<u>m</u> = 0.04		$20^m \pm 20^m$ before moonset

[a] The magnitude of the diameter.

[b] In this column, <u>B</u> means the beginning, <u>M</u> the middle or maximum, and <u>E</u> the end. <u>A</u> means the phase that ibn Yunis calls appearance, and <u>T</u> means the phase that Caussin translates as <u>attouchement</u>. <u>m</u> denotes an instantaneous value of the magnitude of the diameter.

[c] In this column, a notation such as <u>a</u>(α Tau) means the altitude of α Tauri.

[d] Separate statements about the time of the same event.

[e] Alternative readings of the text.

1019 September 16. Source: al-Biruni [1025, p. 261].
al-Biruni made these observations in Ghazni: "When the cut
at the edge of the full moon had become visible," he measured the altitudes of four stars. The stars and the altitudes are: α Aurigae at slightly less than 66° (I use 65°.8
as I did before [Newton, 1972a]), α Canis Majoris at 17°, α
Canis Minoris at 22°, and α Tauri at 63°. I take this to be
what ibn Yunis calls the appearance, with a magnitude of
0.01 of the diameter.

al-Biruni says that he could not observe the end of the
eclipse because a trace of it was still visible when the
moon set behind the mountains. I do not know how much earlier the moon set for him than it would have set behind a
flat horizon, but I assume that the time was $20^m \pm 20^m$ before moonset for a flat horizon, and that the magnitude at
this time was ½ digit of the diameter.

Table V.2 summarizes the data found in these records.
It has the same format as Table V.1. When a record gives
the time of an event in two different ways, such as the
altitude of a star and the time after sunset, I list both
ways. When the reading of a record is uncertain, I give all
the readings that have been suggested.

5.   The Times Derived from the Records of Eclipses

The times of the observations presented in the two
preceding sections pose so many problems that I will deal
with them before continuing with the main narrative. I
start by pointing out two principles I will follow in dealing with the rising and setting of the sun and moon. We are
involved with these times in two separate ways.

First, some of the times are stated either as some
number of unequal hours or as some number of equal hours
from sunrise or sunset. In all these cases, I suspect we
are dealing with calculated times of sunrise or sunset; I
doubt that an astronomer kept a time piece of some sort,
that he could adjust it to keep unequal hours regardless of
the season, and that he started it running at the instant
when the sun disappeared from his sight.

On the contrary, it was standard for Islamic zijes to
contain a table of the rising and setting times of the sun,
which were calculated [Kennedy, 1956] as an exercise in
spherical trigonometry, with the sun being treated as a
point. The astronomers probably took sunrise and sunset
from one of these zijes. Hence, in dealing with times given
with reference to sunrise or sunset, including those given
implicitly this way by means of unequal hours, I find the
times when the center of the sun was on the horizon, neglecting refraction. That is, I take the true altitude of
the center of the sun to be 0°.

Second, we are told sometimes that the sun or moon rose
or set eclipsed, and we are given the value of the magnitude
at this time. Here we have an actual observation, so we
must deal with the apparent altitude of the body in question

instead of its true altitude. However, giving an estimate
of the magnitude implies that all the disc was above the
observer's horizon.

TABLE V.3

TIMING DATA IN ISLAMIC RECORDS OF ECLIPSES

Date	Phase	Object	True altitude °	Greenwich mean time from record h	from Oppolzer[a] h
829 Nov 30	B	Sun	6.876E	4.6013	
	E	Sun	23.964E	6.4471	
854 Feb 16	B			19.3442	19.50
854 Aug 12	B	α Tau	45.484E	0.0090	-0.13
856 Jun 22	B	α Tau	9.406E	0.4225	0.30
866 Jun 16	B			9.5703	
	M			10.7162	
	E			11.9807	
883 Jul 23	M			17.6348	17.10
891 Aug 8	M			10.5731	
901 Jan 23a	M			6.1781	
901 Jan 23b	M			6.1514	
901 Aug 3a	M			0.9873	0.47
901 Aug 3b	M			1.0439	0.47
923 Jun 1	m=0.27	Moon	-0.300E	16.0496	
	M			17.7025	17.43
	E			19.0358	18.85
	E	α Cyg	29.471E	18.9365	18.85
923 Nov 11	M	Sun	7.890E	4.3687	
	E	Sun	19.956E	5.5376	
	E			5.5539	
925 Apr 11	B	β Lib[b]	10.919E	16.8384	16.42
	B			16.3202	16.42
	E	α Lyr	23.964	19.7899	19.82
	E			19.7235	19.82

[a]Given for lunar eclipses only.

[b]This is not the star that appears in the record. See the
accompanying discussion.

TABLE V.3 (continued)

Date	Phase	Object	True altitude °	Greenwich mean time from record h	from Oppolzer[a] h
927 Sep 14	B B̲	αCMa	30.973E	0.9027 0.8862	0.98
928 Aug 18	m=0.42 E	Sun Sun	-0.300E 11.813E	2.5153 3.4987	
929 Jan 27	B	α Boo	17.951E 32.975E 37.979E	20.0830 21.3185 21.7215	21.07
	B̲			20.1701	21.07
933 Nov 5	B B̲	α Boo	14.940E	1.3514 1.2563	1.62 1.62
977 Dec 13	A E̲	Sun Sun	15.442 33.309	6.3014 8.6204	
978 Jun 8	A E̲	Sun Sun	55.989W 25.967W	12.3757 14.7071	
979 May 14	m̲-0.71? E	Moon	-0.300E	16.6805 17.8271	17.80
979 May 28	A m=0.̲46?	Sun Sun	6.367W -0.300W	16.2206 16.7834	
979 Nov 6	B E̲	Moon Moon	64.492E 64.992W	20.0731 23.2930	19.98 23.12
980 May 3	B E̲	Moon	47.652 41.648	c -1.6351 2.5457	-1.27 2.40
981 Apr 22	B E̲	Moon	20.958	1.3913 3.0619	1.55 3.12
981 Oct 16	T̲	Moon	23.964	2.1193	1.80[d]
983 Mar 1	A̲	Moon	65.993 61.991	c 21.0148	21.92[d]
	E	Moon	35.811	25.5452	25.32

[a]Given for lunar eclipses only.

[b]This is not the star that appears in the record. See the accompanying discussion.

[c]Impossible condition.

[d]Time of beginning.

TABLE V.3 (continued)

Date	Phase	Object	True altitude °	Greenwich mean time from record h	from Oppolzer[a] h
985 Jul 20	B	Sun	22.962	14.9414	
	E	Sun	5.858	16.3051	
986 Dec 19	A	Moon	23.964W	2.8511	2.58[d]
	T	Moon	50.487	c	2.58[d]
			28.470	2.4851	
990 Apr 12	B	Moon	37.979	19.6791	20.22[d]
			32.975	19.1246	
	T	Moon	37.979	19.6791	20.22[d]
			32.975	19.1246	
	E	0° Aqr	0.000	23.1504	23.22
993 Aug 20	B	Sun	26.968E	5.5979	
	M	Sun	44.984E	7.0061	
	E	Sun	59.991E	8.2802	
1001 Sep 5	E			17.9749	17.55
1002 Mar 1	B	α Boo	51.987E	21.6022	21.47
		α Boo	11.925E	18.4410	
		α Vir	51.987E	22.6110	21.47
		α Vir	11.925E	18.6604	
	B	α Aur	13.936W	21.6837	21.47
		α Tau	13.936W	20.0516	
	E	α Boo	34.977	28.5827	25.10
		α Vir	34.977	20.5910	
1004 Jan 24	B	Sun	18.452W	13.8488	
		Sun	17.449W	13.9422	
	A	Sun	16.446W	14.0346	
	m=0.25	Sun	14.940W	14.1716	
	m=0.50	Sun	9.911W	14.6169	
	M	Sun	4.836W	15.0509	
1019 Apr 8	m=0.39			1.6175	
1019 Sep 16	A	α Aur	65.793	21.6085	21.70[d]
		α CMa	16.948E	21.6510	21.70
		α CMi	21.960E	21.6220	21.70
		α Tau	62.992E	21.6927	21.70
	m=0.04			24.9778	24.73[e]

[a]Given for lunar eclipses only.

[b]This is not the star that appears in the record. See the accompanying discussion.

[c]Impossible condition.

[d]Time of beginning.

[e]Time at the end.

Thus I assume that observations of this sort happened when the lower limb of the body was on the apparent horizon. This gives the first (or last) instant when the observer could see the total disc, instead of being the first (or last instant) when he could see any of it. In putting the lower limb on the horizon, we allow 34' for refraction, and we use the constant amount of 16' for the semidiameter, whether we are talking about the sun or moon. Thus for these observations we take the true altitude of the body to be -18' or -0°.3.

Table V.3 lists all the timing observations involved in the Islamic records of eclipses, either solar or lunar. The first column gives the date of the eclipse as it is tabulated in Oppolzer [1887]; using this date occasionally requires the use of an hour that is less than 0 or greater than 24. The second column gives the phase of the eclipse that was observed. In this column, B is the phase that the observer called the beginning, M is the phase he called the middle or maximum, and E is the phase he called the end. A is the phase that ibn Yunis calls appearance and T is the phase that his translator rendered as attouchement.

In a few cases, we have a measurement of the time when the magnitude had a particular value. I denote these times by writing "m = 0.27", for example, in the second column. In two instances, for the eclipses of 970 May 14 and 979 May 28, it is not clear whether the magnitude mentioned was that at moonrise and sunset, respectively, or whether the magnitudes stated occurred at some other time. I indicate this uncertainty by putting a question mark after the statement of the magnitude.

When a record specifies a time by giving the altitude of some object, I list this object in the third column of Table V.3, and I list the altitude in the fourth column, after correcting it for refraction by the method described in Section III.4. If the record simply states the hour while giving no details of how the hour was determined, columns 3 and 4 are blank. Finally, the fifth column gives the time deduced from the record, and, for lunar eclipses, the last column gives the time of the appropriate phase taken from Oppolzer [1887]. These times are fairly accurate, so we can use them in deciding between possible readings of the recorded times, but we cannot use them in inferring y.

Let us look first at the lunar eclipse of 929 January 27, which tells us two things. The record says that the height of α Bootis was 18° to the east at the beginning of the eclipse, and that this made the time 5 unequal hours of the night. Newcomb [1875, p. 54] and I [Newton, 1970, p. 232] found that the time deduced from the record is seriously in error and discarded it from our final inferences. I also noted that there is good agreement between the times deduced from the ways of stating the time.

Table V.3 confirms the earlier conclusions. The time when α Bootis had an apparent altitude of 18° east is 20.0830 hours, Greenwich mean time, while 5 unequal hours of the night at the point of observation is 20.1701 hours. However, this time differs from Oppolzer's time by about an hour, and it is clearly wrong. Knobel [1879, p. 339] suggests that the altitude rendered as 18° was originally either 33° or 38°, and both values are tested in Table V.3. It is clear that one of these values is correct and that 18° is a misreading.

From the astronomical calculations, we cannot safely choose between 33° and 38°. However, Knobel prints the forms that the numerals 18, 33, and 38 had, and to my eye 18 is closer to 33 than it is to 38. Hence I take the apparent altitude to be 33° and the time of the observation to be 21.3185 hours, Greenwich mean time.

On this occasion, at least, the time stated by the observer was clearly calculated from the astronomical observation that he gives, and we are probably safe in taking this to be the general practice of careful astronomers. After all, they had no accurate timepieces, so far as I know.† Further, since the stated time is seriously in error while agreeing accurately with the stated altitude, the stated time was derived from the erroneous value 18°. That is, the mistake was not made by a copier or a translator. It was already made by the person who calculated the time, and this person was probably the original observer.

The first time in Table V.3 that causes a problem is the beginning time for the lunar eclipse of 925 April 11. The record in the form we now have it says that the height of α Bootis at the beginning of the eclipse was 11°, and that this time was 55 minutes in unequal hours. However, Newcomb [1875, p. 46] finds that the altitude given for α Bootis is impossible for the beginning of the eclipse, and that its altitude must have been 30° or more. I tried to see if any other bright star might have had an altitude of 11° and concluded that the only likely possibility was β Librae. Thus I changed the identity of the star from α Bootis in Table V.2 to β Librae in Table V.3.

We see that the time deduced by taking the star to be β Librae does not agree well with the astronomer's statement of the time. Hence the star is not likely to have been β Librae. Since the direct statement of the time seems reasonable, I use it. That is, I take the measured time of the beginning to be the equivalent of 16.3202 hours, Greenwich time.

The record of 980 May 3 says that the height of the moon was 47° 40′ at the beginning, but this condition turns out to be impossible. Knobel [1879] suggests that the original reading was 41° 40′. This leads to -1.6351 hours, Greenwich time, which is reasonable and I use it.

---

†I have seen it stated that the water clock had been developed to an accuracy of a minute or so a day by this time in history, but I cannot locate the place and I do not know if the statement is correct. In this case, the time clearly did not come from reading a clock.

The record for 983 March 1 says that the altitude of
the moon at the phase called appearance was 66°, but Knobel
suggests that the reading should be 62°, and I have tried
both in Table V.3. The altitude of 66° turns out to be
impossible. The altitude of 62° was attained, but at a time
about an hour too early. Thus the time of appearance cannot
be used.

The record for 986 December 19 says that the height of
the moon at appearance was 24° and at attouchement was
50° 30'. This much difference in time is unreasonable, and
Knobel suggests that the second altitude should be 28° 30'.
The latter value leads to a reasonable result and I adopt
it. Incidentally, this way of stating the time of attouchement seems unlikely if this time had been calculated from
the time of appearance, as I speculated earlier that it was.
It seems to me that ibn Yunis, if he calculated the time of
attouchement, would first have found the time of appearance
from the observation given and that he would then simply
have subtracted a reasonable amount of time. It seems
unlikely that he would have taken the trouble to calculate
the altitude of the moon at this time and give it, without
giving the time he had already calculated. Thus it looks as
if this phase was actually an observed one on this occasion.

For 990 April 12, ibn Yunis says explicitly that beginning and attouchement mean the same thing, but he again
gives an altitude of the moon for this phase. Again it
appears that he observed this phase somehow. I decided in
Section V.4 to try both 33° and 38° for the altitude of the
moon, and the value of 38°, which is the value in the record, looks to be correct. At the end of the eclipse, the
point at the beginning of Aquarius (called 0° Aqr in the
table) was rising. Since we are clearly dealing with a
calculated point, I take this to be the condition when its
true altitude was 0°. In my earlier work [Newton, 1970, p.
229] I made an accidental error in deriving the time from
this observation and got 21.62 hours. As a result, I
thought there was a mistake in the record and discarded
it. The correct time from Table V.3 is 23.1504 hours, and
there is no problem in using this record.

Now we come to the beginning of the eclipse of 1002
March 1. The record as we have it says that the altitude of
α Bootis was 52° to the east. Knobel says that the star
should be α Virginis and that the altitude should be 12°.
When we try both altitudes for both stars, we find that the
only combination that makes sense is the one given in the
printed form of the record.

The text also gives the beginning of the eclipse by
saying that the altitude of α Aurigae was 14° to the west.
Knobel says that the translator (Caussin) made an error in
translating this record and that the star should be α Tauri.
We see from Table V.3 that again Caussin is right and that
Knobel is wrong.

Finally, the record says that the height of α Bootis was 35° at the end, and Knobel says that the correct identification of the star is α Virginis. Here neither reading makes sense, and we cannot use the end time.

In my earlier work, I did not attempt to use this record because of all the comments that had been made about the errors in it. Actually, it poses only one problem, and that is the observation which gives the end time. We have two measurements of the beginning which agree with each other quite well, and I will use the average.

We have a slight problem with the solar eclipse of 1004 January 24. The record says that the height of the sun at the beginning was 18° 30', but Knobel says we should read this as 17° 30'. As we might expect, these readings give times so close together that we cannot tell which is correct. However, I do not use the time of beginning whenever we also have the time of appearance, so the uncertainty is unimportant.

For a variety of reasons, including some of Knobel's emended readings of the text, we are now able to use considerably more of the Islamic records than were used in the earlier studies. I will make a special study of Knobel's emendations in Section IX.6, where I will find that only two of them are correct. Thus they do not have the value that I originally hoped for them.

Many records state a time in two or more ways. When a record gives a usable measurement of the altitude of some body as well as a statement of the time that the astronomer deduced from that altitude, I use only the altitude itself. (That is, I use the time that I deduce from the measured altitude.) When a record gives several measurements of the altitudes of different bodies for the same event, I use the average of the times I deduce from the measurements.

6.   The Astronomical Tables of al-Khwarizmi

al-Khwarizmi's Zij al-sindhind was prepared sometime about 800, or perhaps a little later. It was translated into Latin, presumably by Adelard of Bath, in the early part of the twelfth century, and H.H. Suter published a critical edition of this translation in 1914 (see Neugebauer [1962, Preface]). Neugebauer, in the place just cited, published a translation of Suter's edition, occasionally supplemented by the use of a manuscript that came to light after Suter's edition. Along with his translation, Neugebauer made a detailed analysis of the astronomical methods behind the tables, but he did not publish the tables themselves.

As I quoted Toomer [1973] in Section V.1, the tables show no "trace of original observation" and no trace of "more than trivial computation" by al-Khwarizmi. Toomer goes on to remark that this is strange because al-Khwarizmi is known to have been acquainted with measurements of the obliquity that lead to the rather accurate value of 23° 33', but he chose to adopt the grossly inaccurate Ptolemaic value

of 23° 51'. Then Toomer further says: "Even more inex-
plicable is why, if he had the Ptolemaic tables available,
he preferred to adopt the less accurate parameters and ob-
scure methods of Hindu astronomy."

I have studied only the tables of the sun and moon, but
for these bodies I am not sure that Toomer's last remark is
entirely correct. Certainly the methods of Hindu astronomy
used for the sun and moon are basically less accurate than
those of Greek astronomy, but they are correspondingly simpler and easier to understand and use. Further, so far as
they go, the parameters of Hindu astronomy seem to be about
as accurate as those of Greek astronomy.

To give an example, Greek astronomers well before the
time of Hipparchus had adopted a model for the motion of the
sun that leads to the following relation between its longitude λ and its mean longitude $\underline{L}$:

$$\lambda = \underline{L} - \tan^{-1}\{[\underline{e} \sin (\underline{L} - \underline{A})]/[1 + \underline{e} \cos (\underline{L} - \underline{A})]\}. \qquad (V.7)$$

In this, the angle $\underline{A}$ is called the longitude of apogee. The
Hindus used a simpler form:

$$\lambda = \underline{L} - \underline{c} \sin (\underline{L} - \underline{A}). \qquad (V.8)$$

Equations V.7 and V.8 are the same through the terms that
are linear in $\underline{e}$, and both agree with modern theory to this
accuracy. When we go to higher-order terms, Equation V.7 is
more accurate, but the difference in accuracy is less than a
minute of arc, which is unimportant for observations made
with the naked eye.

al-Khwarizmi's value of e is [Neugebauer, p. 96]
2° 14', Ptolemy's value is 2° 23', and the value from modern
theory is 1° 58'. The Hindu parameter is more accurate.

al-Khwarizmi uses Equation V.8 for the moon as well as
the sun, although neither Equation V.7 nor V.8 applies at
all well for the moon. We can describe the lunar situation
approximately if we use Equation V.7 and make $\underline{e}$ a function
of the lunar elongation $\underline{D}$; Ptolemy uses a lunar model that
accomplishes this. In this approach, the average value of $\underline{e}$
for the moon [Newton, 1977, p. 107ff] is $6°.29$. It is only
$5°.02$ when the moon is full or new, and it increases to
$7°.56$ when the moon is at either quarter.

al-Khwarizmi, however, takes $\underline{e}$ in Equation V.8 to have
the constant value [Neugebauer, p. 96] of $4°.93$. Ptolemy's
value for the full moon is 5° 1', and the modern value is
the same to this accuracy. It is plausible that the Hindu
value $4°.93$ was derived only from observations of lunar
eclipses, when the moon is always full, and that no attempt
had been made to fit a model to the moon at other phases.
Even so, the Hindu parameter is less accurate than the Greek
one, and the Hindu model is far less accurate for a general
phase of the moon.

For our purposes, only al-Khwarizmi's tables of the
mean longitudes $\underline{L}_S$ and $\underline{L}_M$ of the sun and moon are likely to

be useful. While Neugebauer does not print al-Khwarizmi's tables, he gives the necessary parameters in his analysis of the tables [pages 42, 90, and 92]. From these parameters we get

$$\underline{L}_S = 113°.430\ 0 + 0°.985\ 603\ 512\underline{d},$$

$$\underline{L}_M = 117°.754\ 7 + 13°.176\ 355\ 56\underline{d}. \tag{V.9}$$

In these, $\underline{d}$ is time in days from the fundamental epoch, which is the noon before the beginning of the era of the Hijra, in Arin time. Arin is the modern Ujjain in India. When we transform to Greenwich time, we find that the epoch has the Julian day number 1 948 437.789 4.

When we calculate the length of the year from the first of Equations V.9, we get 365.258 439 days. This is clearly the sidereal year rather than the tropical year, in spite of the fact that the positions clearly refer to the equinox. The modern value of the sidereal year is 365.256 360 days. Ptolemy nowhere states the value of the sidereal year, but he gives some data [Chapter IV.2] from which we can calculate that it equals 365.259 859 days. Again the Hindu parameter is more accurate.

There is a basic contradiction in combining a sidereal mean motion with a position referred to the equinox. If we knew the dates of the basic observations back of the constant terms in Equations V.9, the contradiction would not matter; we would just transform Equation V.9 back to the time of the observations. Unfortunately, we do not know the dates of the observations, and thus we cannot use $\underline{L}_S$ and $\underline{L}_M$ directly with safety.

However, the confusion about the reference position should cancel out when we subtract $\underline{L}_S$ from $\underline{L}_M$ to get $\underline{D}$, just as it did with the Alphonsine tables in Section IV.4. Equations V.9 give us

$$\underline{D} = 4°.324\ 7 + 12°.190\ 752\ 05\underline{d} \tag{V.10}$$

for the lunar elongation. The length of the month derived from Equation V.10 is $29^d.530\ 582$, which is rather accurate. Ptolemy's value is $29^d.530\ 594$ and the modern value is $29^d.530\ 589$. The Greek value is slightly more accurate.

Since the mean motion in Equation V.10 is rather accurate, we do not need to know the date of the observations accurately. We may safely assume that it is before al-Khwarizmi, and it should be sufficiently accurate to take the date to be the era of the Hijra.† Also, while I have carefully refrained from mentioning the matter until now, the epoch $\underline{d} = 0$ is mean noon at Ujjain, and we need to know what mean moon means. Unlike the case of the Hakemite

---

†This is less than two centuries before the probable date of al-Khwarizmi's zij.

tables, we do in fact have al-Khwarizmi's table of the equation of time. Unfortunately, he does the same thing as Ptolemy, which is to deal with time intervals rather than with mean time itself, and I find his discussion of the equation of time rather confusing. The reader can consult Neugebauer [1962, pp. 63ff] for an analysis of al-Khwarizmi's equation of time.

If we ignore the definition of the equation of time used in al-Khwarizmi's tables, we introduce an uncertainty in the fundamental epoch of about 900 seconds (Section V.2 above). Here I do not think we need to worry about this uncertainty. At the rather early stage of astronomy involved in al-Khwarizmi's tables, it is reasonable to say that the error in measuring the elongation of the moon is about equal to its semi-diameter, or about 15'. It takes the elongation about half an hour, 1800 seconds, to change this much, and an error of 900 seconds is unimportant in this context.

Thus I use Equation V.10 at the epoch $d = 0$, which is Julian day 1 948 437.789 4. From Equation I.26, the value of dynamical time when D has the value $4°.324\ 7$ is day 1 948 437.841 5. Hence ΔT is $0^d.052\ 1$, or $4\ 500^s$. From Equation I.15, this gives

$$\underline{y} = -15.7 \pm 6.3. \qquad (V.11)$$

The standard deviation assigned in Equation V.11 is the amount $\underline{y}$ changes if we change ΔT by 1800 seconds.

Instead of analyzing the main body of the Islamic observations here, I will go on with the presentation of data from other cultures. The Islamic observations will be analyzed in Chapter IX.

NOTE ADDED IN PROOF

D. A. King and O. Gingerich (Some astronomical observations from thirteenth-century Egypt, Journal for the History of Astronomy, xiii, pp. 121-128, 1982) give the texts and a translation of records of thirteen observations made either in Qus or in Alexandria between 1273 and 1284. Six of the observations relate to conjunctions (or occultations) of a planet and the moon, and the others relate to conjunctions of planets with each other or with a star. Unfortunately, the records were not made with enough care to let them be useful in this work.

CHAPTER VI

GREEK AND CHINESE OBSERVATIONS BEFORE 600

1.   Records from Late Greek Astronomy

Relatively little Greek astronomy survives except the
work of Ptolemy, and this is true both of the period after
Ptolemy as well as the period before him. However, there is
a great difference between the two periods. We know that
much good work was done in Greek astronomy before Ptolemy,
and that most of it has been lost. We believe that little
good work was done after Ptolemy, so there was not much from
the late period either to be lost or to survive.

As I point out in Appendix IV, the great period of
Greek scholarship was over by -200, and we have only a few
isolated examples of good scholarship after this time.
Nonetheless, Greek scholarship lingered on until the reign
of Justinian, who was the emperor of the Byzantine empire
from 527 to 565. It is probably fair to the intellectual
standards of the time to point out that we have no work on
astronomy dating from the reign of Justinian, so far as I
know. We have only some notes written in the margins of a
copy of Ptolemy's Syntaxis. Luckily, a scribe used this
copy to make other copies, and he did not care that the
marginal material was not part of the original; he copied
everything before him, marginal notes and all. Thus the
marginal notes have been preserved for us.

This copy belonged to Heliodorus who, along with his
brother Ammonius, was a student of Proclus in the academy at
Athens for some time, apparently including the year 475.
Proclus was a vigorous opponent of Christianity, and he was
probably the last significant figure who upheld the school
of philosophy called NeoPlatonism. Pauly-Wissowa [1894,
volume 1, p. 415 and volume 8, p. 18] summarizes the scholarly attainments of Ammonius and Heliodorus as succinctly as
I can imagine by saying that Heliodorus was more undistin-
guished than his brother.

After their student days in Athens, Ammonius and Heliodorus went to teach in Alexandria at some time and were
there at least during the years 498 to 510. One of their
students named Damascius subsequently went to Athens and was
the head of the academy there [Pauly-Wissowa, volume 2, p.
847] when Justinian ordered it closed in 529.

In his copy of the Syntaxis, Heliodorus recorded an
astronomical observation made by Proclus in Athens in 475
and six observations that were made in Alexandria from 498
to 510. Of these six, as well as we can judge by the language, five were made by Heliodorus alone and the other was
made jointly with his brother. I gave translations of all
seven observations in an earlier work [Newton, 1976, pp. 210 -
213], along with citations of the original sources. Here I
repeat only the observations that involve the moon.

475 November 18. The moon occulted Venus on this date. As seen from Athens, the apparent longitude of the moon was 283°, and it was 48° from the sun. If Heliodorus has represented adequately what Proclus observed, Proclus was not a good observing astronomer, whatever may have been his qualifications as a philosopher. He does not even give the times when the occultation began and ended, and, since he gives the positions only to an integer degree, we cannot reconstruct the times by calculation. By calculation, I concluded earlier [Newton, 1976, p. 413] that the occultation was centered at about sunset in Athens, but it would be reasoning in a circle to use this conclusion in studying the secular accelerations. Hence I do not use this record.

503 February 21. Heliodorus and his "most beloved brother" made this observation jointly in Alexandria. At the first hour of the night, the moon was west of Saturn. When the moon had passed Saturn, they measured the time with an astrolabe and found that it was 5 3/4 unequal hours (presumably after sunset). They concluded that the conjunction, as seen at Alexandria, was at 5 1/8 hours, presumably unequal hours after sunset. They judged that Saturn bisected the illuminated curve of the moon.

We should note that this record is of higher quality than the record of the observation by Proclus. However, we do not know how well Heliodorus preserved the actual work of Proclus, so we should not compare the observers rashly.

We cannot use the statement that Saturn bisected the curve of the moon in finding the accelerations, but we can use it to judge the quality of the observation. I do this in Section VIII.7, where I find that the quality is rather poor.

509 March 11. This evening, in Alexandria, α Tauri was west of the bright arc of the moon by at most 6 digits. In this context, 6 digits means 0°.5. We have two difficulties in using this record. One concerns the time and the other concerns what is meant by the bright arc of the moon.

To start with, the moon was between 4 and 5 days old at the time of the observation, so it set before midnight. In other records that simply record an observation as being made in the evening, the time seems to mean as soon after sunset as the objects could be seen and identified. The moon can be seen before sunset, but it rarely is unless the observer figures out where to look, at least in my experience. The magnitude of α Tauri is 1.06 [Becvar, 1964], so the star itself could probably be seen fairly soon after sunset. However, the observer should wait until he can see some other stars in the vicinity in order to be sure of the identification. I think it is reasonable to take the time to be 45 minutes after sunset.

The other problem concerns where the bright arc of the moon seems to be to the unaided eye. This problem already came up in Sections II.4 and IV.3. As I have already noted,

Fotheringham [1915, p. 388] assumes that a star appears occulted when it is within 3' of the bright limb of the moon,
but he gives no observations to support his assumption. I
studied Tycho's measurements of the lunar diameter in Section II.4 and found (Equation II.3) that he regularly measured it to be $1'.52 \pm 0'.57$ too large. This amounts to
displacing the bright limb by $0'.76$, much smaller than the
amount Fotheringham uses. I continue with the choice I made
in Section IV.10, which is to use the value $0.'76 \pm 0'.28$
for the displacement of the bright limb. This is $0°.013
\pm 0°.005$. For an observation this old, the limb correction
is not important.

Thus, if α Tauri was $0°.5$ west of the bright arc as it
appeared to Heliodorus, I take it to be $0°.513$ west of the
true limb.

Going backward in time, we come to an astronomer named
Theon who worked in Alexandria in the last half of the
fourth century.† This Theon is often called Theon of Alexandria, apparently in order not to confuse him with another
astronomer from Alexandria named Theon, who was a contemporary of Ptolemy and whom Ptolemy credits with certain
observations of Mercury and Venus. Theon of Alexandria
wrote a considerable amount of commentary on the Syntaxis,
some of which has survived.

We date Theon of Alexandria partly by the fact that his
surviving work contains an observation of the solar eclipse
of 364 June 16 that he made, presumably in Alexandria.
Fotheringham [1920] gives the following translation of the
record:‡ "And moreover we observed with the greatest certainty the time of the beginning of contact, reckoned by
civil and apparent time as 2 5/6 equinoctial hours after
midday, and the time of the middle of the eclipse as 3 4/5
hours, and the time of complete restoration as 4 1/2 hours
approximately after the said midday . ." Theon has already
given the date of the eclipse in the Egyptian calendar and
he goes on to repeat the month and day.

---

†The work of Theon and the work of Heliodorus show that
Alexandria continued to be a center of whatever ancient
Greek culture was left down to the essential extinction of
classical culture under Justinian around 530. This suggests, among other things, that the famous library at
Alexandria still survived, probably in an attenuated form,
and that it was not destroyed by fire during Caesar's
campaign there. See Appendix IV.

‡Fotheringham considers records of this and of ten other
solar eclipses found in ancient writing. Out of these elev-
en, only the records of 364 June 16, -309 August 15, and
-430 August 3 are both valid and datable observations. The
other records are either fictional, undatable, or both.
Ironically, Fotheringham rejects the only three valid observations in his sample, on the basis that the results
from them are not compatible with the other results.

I do not know how to interpret the fact that Theon measured the beginning "with the greatest certainty" but that he measured the end only "approximately". Unless he were hampered by clouds near the end, I doubt that one measurement was really more accurate than the other, and I will give equal weights to both the beginning and the end. The middle time, namely $3^h 48^m$, differs by 8^m from the average of the beginning and end, which is $3^h 40^m$. Hence, the middle time is an independent measurement, but I do not see how one can judge the middle, which I assume really means the maximum, with as much precision as the beginning or end. Hence I will give the middle time only a fourth of the weight of the other measurements.

This observation by Theon, and the ones recorded by Heliodorus, are the only observations by Greek astronomers I know of after Ptolemy. There is a statement by Censorinus [238] about the position of the sun relative to Sirius, but I believe it embodies a conventional statement instead of being an actual observation [Newton, 1976, pp. 204ff].

2. Records from Ptolemy's Syntaxis

It is interesting to see how difficult it is to overcome the viewpoints one absorbs during one's education. I was taught that Ptolemy's Syntaxis [Ptolemy, ca. 142] represents the pinnacle of ancient astronomy, and such a statement has been printed several times even during the past decade. Thus, when I discovered [Newton, 1970, Section II.2] that Ptolemy simply made up his observations of the sun in order to "confirm" the accuracy of older tables of the sun, I mentally wrote this off as a minor aberration in a great man that probably has an innocent explanation (if one could only think of it). I even suggested that Ptolemy had an assistant charged with doing the actual observing who systematically cheated on his employer.

Later, I discovered that Delambre [1819, p. lxviii] and Britton [1967] had independently made the same discovery about the solar observations. Britton could make this discovery independently of Delambre, and I could later make it independently of both of them, because the earlier discoveries had been totally ignored in the literature,† and it was only by accident that I discovered these references. This made me wonder some about the care with which students of the history of astronomy studied their own literature, but it naturally had nothing to do with Ptolemy himself. Even when I discovered [Newton, 1972b] that Ptolemy fabricated his measurements of the obliquity of the ecliptic, I still did not think to question the basic stature and integrity of his work.

---

†Britton's 1967 work was his doctoral dissertation. In a later paper dealing with some of Ptolemy's solar observations [Britton, 1969], even Britton failed to mention this part of his own work.

Only when I realized that Ptolemy also fabricated the "observation" from which he got the lunar parallax [Syntaxis, Chapter V.13] did it occur to me that we might be dealing with a massive fabrication of data rather than with a work of outstanding scholarly merit. At this point I decided to make a systematic analysis of all the observations that Ptolemy claims to have made himself. My conclusion [Newton, 1976, Chapter V] was that Ptolemy, with one exception,† fabricated all the observations that he claims to have made.

In this same work I also considered the question of whether Ptolemy might have fabricated the observations of earlier astronomers that he reports, and I concluded that he did not, with at most minor exceptions. My main argument [Newton, 1976, pp. 156ff] was that the work of the earlier astronomers was still widely available in Ptolemy's time, and that he would have run the risk of instant detection if he had dared to alter their reported results. Only when I had written about a third of my analysis of the Syntaxis did I realize [Newton, 1977] that even this conclusion was too optimistic. Ptolemy in fact fabricated a large fraction of the observations that he attributes to earlier astronomers.

The fact that Ptolemy did fabricate observations attributed to earlier astronomers suggests a great deal about the status of Greek scholarship at the time, including the status of the library at Alexandria. For one thing, Alexandria, where Ptolemy worked, was and would be for another four centuries a center for whatever Greek scholarship still remained, as I said above. For another thing, a study of most of the fabricated observations indicates that Ptolemy did not make them up out of the whole cloth. Instead, he probably started from actual records and altered the numbers in them to suit his purposes, in most cases.‡ However, the fact that he dared alter the numbers suggests that few scholars were likely to consult the original records and thus detect his alteration.‡ Finally, his activity suggests that the library at Alexandria was still functioning; this library is the most likely repository for the records that Ptolemy

---

†The exception is connected with the declinations of eighteen stars that Ptolemy [Syntaxis, Chapter VII.3] claims to have measured in order to determine the precession of the equinoxes. He uses the declinations of only six stars in his actual analysis, and all these are fabricated. The declinations of the other twelve stars are genuine measurements, but he uses them only as "camouflage" to make it appear that the six values used were selected at random.

‡I suspect that a few of the observations were made up in their entirety, and I will take up some of these cases in a moment.

‡To be sure, some "scholars" even today fabricate their data. However, I do not remember any "scholars" other than Ptolemy and Copernicus who fabricated data that were allegedly received from earlier scholars. See Section III.2.

needed to start his process of fabrication.

In my first work on the secular accelerations [Newton, 1970], I used the timed observations from the Syntaxis except for the solar ones. For the measurements of the magnitudes of partial lunar eclipses, I compared [pp. 217ff] the scatter of the measurements from the Syntaxis with the scatter of the measurements from other sources, and I was mildly puzzled to find that the scatter of those from the Syntaxis was only about half of that from other sources. This led me to suggest that Ptolemy might have selected these measurements on some unknown basis, but I also pointed out that the result could just be a statistical accident arising with such a small sample.

Now, of course, we realize why the scatter of the results is so small: The results were fabricated to fit his lunar theory. Further, while Ptolemy's theory overall is not particularly accurate, its accuracy depends strongly upon the phase of the moon, and it is rather accurate for the full moon, the phase at which lunar eclipses occur. Thus the main error in Ptolemy's eclipse data comes from the rounding he used to make his fabricated values look like plausible results of observation.

It is clear that we must not use the lunar observations that Ptolemy claims to have made himself, but we must still ask whether we may safely use any of the observations attributed to other astronomers. I have summarized all the observations attributed to other astronomers in Table XIII.2 of an earlier work [Newton, 1977, pp. 344 - 346]. I count 26 observations that involve the moon. Of these, 20 are definitely fabricated. I cannot prove fabrication in 3 other cases, but the circumstances are suspicious and I do not consider using them. This leaves 3, which I label "may be genuine" in the table. These are all observations of lunar eclipses. The eclipses of -501 November 9 and -490 April 25 were observed in Babylon, and the eclipse of 125 April 5 was observed in Alexandria.†

When I wrote that these three records may be genuine, I meant that I had found no astronomical evidence that casts doubt upon their validity. However, as I explained later [Newton, 1977, Appendix C], there is what we may call

----

†Pedersen [1974, pp. 408 - 422] gives a list of all 94 observations in the Syntaxis that can be dated, at least within a day or so. For the eclipse of 125 April 5, he lists the observer as "Perhaps Ptolemy". In the record of this observation [Syntaxis, Chapter IV.9], Ptolemy gives no clue to the observer. Further, the date is about two years earlier than any other observation that Ptolemy attributes either to Theon or to himself. Since Ptolemy does not claim that he made this observation, we cannot put it among the observations that he does claim to have made. My suspicion is that he meant to attribute it to Theon but forgot to.

stylistic evidence which indicates that the two eclipses
from Babylon were fabricated and furthermore that they were
fabricated in their entirety. The evidence deals with the
way that Ptolemy states the dates of the observations.

When Ptolemy presents an observation that he attributes
to another astronomer, his standard practice is first to
give the full date in the calendar that the other astronomer
used, and then to give the equivalent date in the Egyptian
calendar, which we may translate without ambiguity into the
Julian or Gregorian calendar. This is true of all observations except those that were allegedly made in Babylon.
Here, he never states any part of the Babylonian date except
the year, and he then immediately gives the full date in the
Egyptian calendar. This is exactly what he would have to do
if he made up the Babylonian "observations" in their entirety, and I can think of no other reason for this practice.

The explanation of this practice lies in the nature of
the Babylonian calendar during the relevant period of time,
which is quite different in principle from modern calendars.
If we are given the year, month, and day of an event in the
Gregorian calendar, the Julian calendar, the Egyptian calendar, and many other calendars, we can immediately calculate the Julian day number of the event. This is not true
of a date given in the Babylonian calendar at the period in
question.

In the Babylonian calendar, the first day of the month
was the first day on which the moon was seen in the western
sky after its period of invisibility.† Thus, on the average, the Babylonian month equalled the true lunar month of
29.530 6 days. Now 12 months of this length equal about
354.4 days, which is about 11 days less than the tropical
year. However, the Babylonian year was tied to the agricul-
tural seasons, so the year had to equal the tropical year on
the average. This means that some years had to have 12
months while others had 13 months. At a late period in
Babylonian chronology, rules were developed to determine the
number of months in each year (see Section VII.6). At the
period we are talking about, the number of months in each
year was still determined by direct observation of the position of the sun. This, in effect, means that the first day
of the year, as well as the first day of the month, was
determined by direct observation.

Suppose that we know, say, the Julian day number of
some event at a time when both the beginning of each year
and the beginning of each month were determined by observation. Unless the event comes too close to the possible
beginning of the year, we can determine the Babylonian year

----

†In this discussion, the statements I make are true only on
the average. For a more detailed description of the Babylonian calendar, I refer the reader to Parker and Dubberstein [1956].

of the event, but we cannot determine either the month or the day of the month.†

Now suppose that Ptolemy calculated the date, in the Egyptian calendar, of some lunar eclipse that had a property he wanted, such as the moon's being eclipsed by 2 digits on the south side when it was near apogee. From his list of correspondences between the Babylonian and Egyptian years,‡ he could then assign the Babylonian year, but, as I just said, he could not assign the month and day. Thus, all he could say is just what he does say, namely that the eclipse came in year $\underline{Y}_B$ of the Babylonian calendar, on the date with year $\underline{Y}_E$ in the Egyptian calendar, on day $\underline{D}$ of Egyptian month $\underline{M}$.

This makes it unlikely that Ptolemy took actual Babylonian records and altered the numbers in them. If he had, he would presumably have followed the same practice that he did with other records. The record of the eclipse of -522 July 16 is particularly informative, because it happens that we have an original cuneiform record of this eclipse. The cuneiform record says that the eclipse came in the 7th year of Kambyses, during the night of the 14th of the month Duzu. If Ptolemy followed his usual practice, he would start by giving this date in full, and then give the Egyptian date. Since he does not do this for any Babylonian observation, he probably did not have access to Babylonian records.‡ We should remember that the ability to read cuneiform was lost sometime in antiquity, perhaps about the time of Ptolemy, and that it was not recovered until the last century.

---

†Since the moon moves so rapidly, we can calculate the first visibility of the new moon with an uncertainty that does not usually exceed a day. The biggest uncertainty comes in the month. When an extra month was added, it was not always added at the same place in the year. It was sometimes added at the end, sometimes between the sixth and seventh months, and occasionally at other places in the year. (See Section VII.6). Further, we do not know what solar event was used to regulate the need for an added month. Some scholars think the rule was that the autumnal equinox must come in the seventh month, others think the first visibility of Sirius in the summer had to come in the fourth month, and so on. If the latter rule was correct, the average length of the Babylonian year was the sidereal rather than the tropical year, and in fact the only attested Babylonian value for the length of the year [Neugebauer, 1955, volume 1, p. 70] is clearly sidereal.

‡This is the list that is often called Ptolemy's king list, since it gives the Egyptian years of the reigns of the Babylonian monarchs.

‡There is independent confirmation of this conclusion in Section VII.1. Also see Section VII.7.

Thus, even though there is no astronomical evidence to say that the records of -501 November 9 and -490 April 25 were fabricated, there is other evidence to this effect, and we should not use these records. This leaves only the eclipse of 125 April 5. Since this record occurs in a suspicious context, the conservative course is to ignore it, even though we have no evidence that it was fabricated. In other words, if a person wants to use this record, the burden of proof should be on him that the record is valid. We should not have to prove that it is invalid in order to ignore it.

Thus, of all the observations in the Syntaxis that involve the moon, there is not one that we may safely use in astronomical research. In fact, there is not a single surviving observation of the moon from all of Greek astronomy before 364 June 16 that is valid.† By an astronomical equivalent of Gresham's Law, Ptolemy's fabricated astronomy has driven out valid Greek astronomy.

This fact shows what is perhaps the most tragic consequence of Ptolemy's fabrications: They have caused us to lose most of the valid work of the earlier Greek astronomers. And let no one doubt that there was a large corpus of earlier observations to be lost. We must conclude that there was a large body of earlier observations whether we accept Ptolemy as a valid source or not.

Let us first accept him as a valid source. He tells us that many of the observations of the moon that he uses were made by Hipparchus, and it is unlikely that Ptolemy presents all of Hipparchus's observations. Ptolemy uses many positions of the stars measured by Timocharis, Aristyllus, and Hipparchus, and it is again unlikely that he presents all of them. He uses many observations of the planets made by the pre-Hipparchan astronomers Dionysios and Timocharis, and by one or more unnamed astronomers in Alexandria, and he then says with regard to Hipparchus [Chapter IX.2]: "And so I consider Hipparchus to have been most zealous after the truth, . . especially because of his having left us more examples of accurate observations than he ever got from his predecessors." In other words, Hipparchus more than doubled the corpus of accurate Greek astronomical observations.

Now let us be more realistic and accept the fact that Ptolemy cannot be accepted as a valid source. In this case, we know that most of the observations in the Syntaxis have been derived with the use of the parameters found there, and not the other way around, as matters should be. However, many of the parameters are reasonably accurate, so they must have been derived from genuine observations. We know that one of them was derived from Babylonian sources,‡ but many

---

†This does not refer to records of solar eclipses that were seen by laymen.

‡This is the length of the month, which I will come back to in a moment.

and perhaps most of them could not have been. The reason
for this statement is that the methods of Babylonian astronomy are quite different from those of Greek astronomy,† and
many parameters in Greek theory have no Babylonian counterparts, nor do they have known counterparts in any other part
of the ancient world. Hence, they must have been derived
from observation by the Greeks, and we have lost the observations they used to do this.

Further, in at least two instances, Hipparchus deduced
from observations, now lost, parameters more accurate than
the ones Ptolemy passes on to us. I have thought of only
one reason why Ptolemy would reject these parameters and go
back to more primitive values.

The first instance concerns the length of the month.
Ptolemy [Syntaxis, Chapter IV.2] says that Hipparchus combined his observations with "Chaldean" ones and found that
4 267 months take 126 007 days plus 1 equal hour. These
values lead to 29;31,50,8,9 (= 29.530 593) days for the
length of the month. However, Ptolemy says that they lead
to 29;31,50,8,20 (= 29.530 594) days, which is an attested
Babylonian value [Neugebauer, 1955, volume 1, p. 78] that is
older than Hipparchus. Ptolemy should have expected Hipparchus's value to be more accurate, which it is (the correct
value is about 29.530 589 days), and he should have adopted
it. I suspect that there were already lengthy tables in
existence based upon the Babylonian month, and Ptolemy did
not want to go to the effort of calculating new tables.
However, it still seems odd that he did not adopt tables
based upon Hipparchus's length of the month; perhaps no one
had calculated such tables.‡

The second instance concerns the obliquity of the
ecliptic. Ptolemy claims [Syntaxis, Chapter I.12] that he
measured the angle between the tropics several times, and
that he always found it to lie between 47;40 and 47;45 degrees. Hence, he confirms Eratosthenes' value for the obliquity, which was 23;51,20 degrees; Eratosthenes' value dates
to about -250. The correct value in Ptolemy's time was
about 23;41 degrees. Turning to Hipparchus, he used 23;55
in his earliest work [Rawlins, 1982], but he later used

----

†I refer the reader to Neugebauer [1955] for a thorough
 discussion of the methods of Babylonian astronomy.

‡In research I have done since writing this, and which I am
 preparing for publication, I have discovered that the relation 4 267 months = 126 007 days + 1 hour was almost surely
 not a measured one. Instead, it was probably obtained by
 multiplication up from a shorter interval that was measured. We do not in fact know what value Hipparchus used
 for the length of the month. I still think it likely that
 Ptolemy adopted the Babylonian month in order to use existing tables, not because it was Hipparchus's value. It is
 also possible that Hipparchus himself adopted the Egyptian
 month, even though his data lead to a slightly different
 value.

23;40 degrees. Yet Ptolemy adopts the value of Eratosthenes, which is older by a century and which should have been less accurate than the value Hipparchus used in his mature work.

Here I feel sure that the explanation comes from tables. The length of the longest day of the year at a particular place was an important geographical parameter in early Greek science, probably more important than the latitude. In Chapter II.6 of the Syntaxis, Ptolemy tabulates the latitude of a point as a function of the longest day of the year at that point, and the tabular values are clearly based upon taking 23;51,20 for the obliquity.† I have shown [Newton, 1983, Appendix D] that this table is far older than Ptolemy and that it may well date back to the time of Eratosthenes.

There is another table in the Syntaxis [Chapter I.15] that also depends critically upon the obliquity. This is a table giving the declination of the sun as a function of its longitude. In work that I am preparing for publication, I will show that this table too is much older than Ptolemy, but not as old as the table mentioned in the preceding paragraph.‡ Thus it seems clear that Ptolemy adopted an archaic value of the obliquity in order to reproduce archaic tables based upon that value. This still leaves the question of why he did not use tables based upon 23;40 degrees. We know [Rawlins, 1982] that there were some pre-Ptolemaic tables based upon this value, but they may not have included all of the functions Ptolemy wanted to present.

3. A Summary of the Greek Observations

Greek astronomy was the mother of modern astronomy, and it was based upon a body of observations that must have been unequalled in accuracy up to that time. In spite of this, there are no surviving observations of the moon from the period when Greek astronomy was at the height of its achievements in both observational and theoretical astron-

---

†This is a computed table in which refraction is neglected and the sun is treated as a point. Thus it differs from a modern table of the same quantities, quite aside from the value used for the obliquity.

‡All the tables in the Syntaxis that I have examined up to the present (May 1983) prove to be older than Ptolemy, with the possible exception of the table of chords [Syntaxis, Chapter I.11]. However, Ptolemy gives no indication that he has taken these tables over from earlier authors and that he has not even taken the trouble to recalculate them using more accurate methods. This does not accord with the praise given him by many writers. See, for example, the statement by Pedersen [1974, p. 13]: "Because he always acknowledged his debts to earlier scientists his own results stand out clearly, . . ." On the contrary, he freely appropriates the work of other scientists as his own throughout the Syntaxis, with no acknowledgment.

omy. The only valid observations of the moon we have come from the very end of the ancient Greek culture, when it had lost almost all of its vitality. One of these valid observations is 2 centuries later than Ptolemy and the others are 3 ½ centuries later. Table VI.1 summarizes the remaining valid observations, and it is unlikely that we will ever discover any more.

TABLE VI.1

SURVIVING GREEK OBSERVATIONS INVOLVING THE MOON

Date	Time[a]	Event
364 Jun 16	2 5/6 hours[b]	Solar eclipse, beginning
	3 4/5 hours[b]	Middle
	4 1/2 hours[b]	End
503 Feb 21	5 1/8 unequal hours	Conjunction of moon and Saturn[c]
509 Mar 11	45 minutes[d]	α Tauri 0°.513 west of lunar limb

[a]All the observations were made in Alexandria.

[b]After apparent noon.

[c]Saturn seemed to bisect the moon.

[d]After sunset.

4. A Chinese Record from the Year -88

I do not know of any Chinese records before the Tarng dynasty that give quantitative statements about the magnitude of an eclipse, and I know of only one record that gives a quantitative statement about the time of one. This is a record of the solar eclipse of -88 September 29.

The record says [Dubs, 1938]: "It was partial, like a hook. In the late afternoon the lower part of the sun was eclipsed from the northwest. In the late afternoon the eclipse was over." I noted this record in volume 1 (page 163), but I did not use it there, since the study of eclipse times and magnitudes is the subject of this volume.

To the modern eye, it looks as if this record has only a vague statement of the time of the eclipse. However, in the terminology used by the translator (Dubs) of this record, "late afternoon" has a quantitative meaning. The Chinese, like the Babylonians, used a twelfth of a day as the basic unit for giving the time of day. A Chinese "dou-

ble-hour" began and ended at an hour that has an odd number
in our system of denoting hours. Dubs uses the term "late
afternoon" to denote the particular double-hour that began
at 15 hours (local time) and ended at 17 hours.

Thus, according to the record, both the middle of the
eclipse and its end came between 15 and 17 hours, in what we
may take to be local apparent time. To anticipate the calculated results, the time from the middle of the eclipse to
its end was about 1.1 hours in the part of China that is in
question. I will assume that this interval of 1.1 hours is
centered about the middle of the double-hour. That is, the
time that is midway between the middle and the end of the
eclipse is taken to be 16 hours, local apparent time.

If we assume that there was no error in determining the
time, this time is uncertain by a maximum of 0.45 hours.
For this maximum error, we should assign a standard deviation of $0.45/\sqrt{2} = 0.37$ hours. Since there could also have
been an error in measuring the time itself, I will be conservative and take the standard deviation in the time to be
0.45 hours.

This value of the standard deviation is based upon the
assumption that we know where the observation was made.
However, the observation could have been made at any place
where the Chinese emperors of the time maintained observatories. We must take this to be (volume 1, p. 141) anywhere
in the rectangle whose center is at latitude 34°.06 and
longitude 113°.51; the rectangle extends 1°.15 in latitude
on either side of the center and 3°.57 on either side in
longitude.

I have calculated the circumstances of the eclipse for
the center of this rectangle as well as for the four corners. The result is that we should use the time calculated
for the center, with a standard deviation of 0.22 hours
introduced because of the uncertainty in the place of observation. This has a small effect compared with the uncertainty in the time. When we combine this with the earlier
contribution of 0.45 hours, the resulting standard deviation
is almost exactly 0.5 hours.

BABYLONIAN OBSERVATIONS

1.   The Nature of Babylonian Astronomy

     Babylonian astronomy in the sense of a mere catalogue
of observations goes back as far as -1500 or earlier. By
the period that concerns us, it had developed into a systematic pattern of observations that were probably made on a
continuing basis, and into rather complex methods of calculating the positions of the sun, moon, and planets. In
spite of its age, the article by Sachs [1948] is probably
still the standard work on the observational texts, and the
mammoth work by Neugebauer [1955] is the standard work on
the computational aspects of Babylonian astronomy.

     Neugebauer [1957, Chapter V] has given a fairly short
description (about 40 pages) of Babylonian astronomy aimed
at the general reader, and I have given a shorter description [Newton, 1976, pp. 97 - 110] aimed at the same reader.
Both sources give references to the earlier literature.
Here I give an even shorter summary, which merely tries to
describe the usual situation; there may be many exceptions
to any general remark I make.

     In Greek astronomy, except perhaps in its earliest
parts, the goal was to set up a coordinate system in the
heavens, to observe the positions of the sun, moon, planets,
and stars in this system, and to develop theories for calculating these positions as functions of time. In the Babylonian astronomy of the period that concerns us, the goal
was different. For the moon, the Babylonian astronomers
were primarily interested in observing and calculating the
times of the full moon, of the first and last visibility
around the times of the new moon, and the occurrence of
eclipses. For the planets, they were primarily interested
in similar "synodic" phenomena, such as the last time a
planet could be seen as it approached the sun and the first
time it could be seen after it had passed the sun.

     In addition, the Babylonians used a set of stars that
are usually called the "normal stars" in the literature.
This is a set of about 35 stars (the authorities disagree
about the exact number) lying near the ecliptic, but whose
principles of selection are puzzling. They are not the 35
brightest stars near the ecliptic; their magnitudes range up
to 4.73 [Newton, 1976, p. 87]. Further, they are not distributed uniformly around the ecliptic; there is a gap of
about 60° between δ Capricorni and η Piscium, neither of
which is particularly bright. The Babylonian astronomers
frequently recorded the passage of the moon and planets past
one of the normal stars. This, of course, amounts to a
measurement of the actual position of the moon or a planet,
and it is thus contrary to the goal mentioned in the preceding paragraph. In fact, we have more observations that
relate the moon and planets to the normal stars than we have
of the synodic phenomena. However, the synodic phenomena
are the principal interest of the computational astronomy.

In Chapter IX.2 of the <u>Syntaxis</u>, Ptolemy writes that
the best sustained sequences of the old planetary observations were mostly of the first and last visibilities and of
the stationary points.† He remarks correctly that such
observations are not good ones to build theories of the
planets on. He then goes on to say that observations of the
angle between a planet and a star are also poor observations
when the angle is large, and his reasons for saying this are
also valid.

I do not believe that Ptolemy could have written these
passages if he had been directly acquainted with the Babylonian observations of the moon and planets, because neither
statement applies to those observations. As I have already
noted, more Babylonian observations give the position of the
moon or a planet than give the various synodic conditions,
so his first statement is not correct for the Babylonian
data. The Babylonian observations do, it is true, give a
position by giving the distance of the moon or a planet from
a star, but the distance is typically 5°, and it is rarely
as large as 10°. Ptolemy's criticisms do not apply to
angles as small as these, and thus neither of Ptolemy's
statements applies to the Babylonian observations.

To be specific, I found [<u>Newton</u>, 1976] the Babylonian
observations of both the moon and the planets to be useful
in studying a certain problem in modern astronomy, and Ptolemy should have found them useful in ancient astronomy. I
showed in Section VI.2 above that Ptolemy probably made up
his "Babylonian" observations of eclipses out of whole
cloth. He probably did not have access to genuine Babylonian observations, in which he simply altered some key numbers. We now have independent confirmation of this conclusion.

Returning to Babylonian astronomy itself, the emphasis
was on the synodic phenomena for both the moon and planets,
even though measurements of position outweigh measurements
of synodic phenomena. Perhaps the reason is simply that
there are more opportunities to measure position than there
are to measure synodic phenomena.‡ The Babylonians developed elaborate methods for calculating the times of the
synodic phenomena, and they did so by purely computational
schemes. That is, they did not develop geometrical models
of celestial motion and work out the consequences, as the

----

†A planet travels from west to east through the stars most
  of the time, but at set intervals it comes to a stop, travels westward for awhile, and then stops again before resum-
  ing its usual eastward direction. The places where it
  stops are called its stationary points.

‡This is true for the moon and all the planets except Mercury. Because Mercury is always close to the sun, and because its synodic phenomena recur so often, there are many
  opportunities to measure the synodic phenomena and few opportunities to measure its position when it is well removed
  from an important synodic position.

Greeks did. Nor did they have physical theories of celestial motions, as we have today. The closest modern analogy I can think of to their astronomical methods is the method of constructing tide tables.

Suppose we want to calculate the times of high and low tides at, say, Atlantic City, New Jersey every day. From the geophysical point of view, the times depend upon the self-gravitation of the earth and oceans, the rotation of the earth, the gravitational attractions of the sun and moon, the detailed shape of the coastlines, and probably other matters I have forgotten to mention. To calculate the times of the tides from first principles is a formidable job and this is not the way it is done.† Instead, the tide tables are calculated with no reference to physical theory, for any place that wants them.

The raw material is a set of measured heights of the tide as a function of time extending over a period of many years. When we look at the measured heights, the first thing we notice is that the height varies sinusoidally with a period of half a lunar day (at most places). From a long run of data, we find the amplitude of this variation and subtract the variation from the observations. When we examine the residuals, we find that they have another strong periodic component, say one with a period of half a solar day. We fit this to the data, subtract it out, look for the most important component in the new residuals, and so on, as far as we are justified by the accuracy of the data and the cost of the process. Then, with all these periodic components known, we simply add them up for any desired time. No physical theory is explicitly involved in this, although theory may suggest the periods to look for.

The Babylonian astronomers proceeded in the same basic way to calculate the times when Mars, say, is in opposition to the sun. The raw material available to the Babylonians may have been a set of days when Mars rose just as the sun was setting, and of the position of Mars at these times. We know that both our year and the Martian year affect these times, but the Babylonian astronomers had to find out the hard way that both the year and another important period of about 2.1 years are involved in the data. After they learned this, they calculated periodic functions with these periods and subtracted them from the data. They then identified other less important periods and continued just as we do with the tide tables.

One interesting point should be made. When we think of a periodic function, we almost automatically think of a sinusoidal function. The Babylonians did not know these

---

†With the aid of electronic computers, some theoretical tidal calculations of moderate accuracy are beginning to be made. However, so far as I am aware, we are still not at the place where we can make routine calculations of tide tables for all important points on the coast from physical theory.

functions, but they used two other types of periodic function. One is the periodic step function. The fundamental periodic step function of x (with unit period) equals +1 for x from 0 to 0.5 and -1 for x for 0.5 to 1. The other function the Babylonians used is the periodic zigzag function. The standard function of this sort increases linearly from 0 to 1 as x changes from 0 to 0.25. It then decreases linearly to -1 as x moves on to 0.75. Finally, it comes back linearly to 0 as x increases to 1.

There are two main systems of Babylonian astronomy, called A and B. System A uses periodic step functions and System B uses periodic zigzag functions. The reader who is interested in studying these systems should start by consulting Neugebauer [1955].

2. Babylonian Measurements of Position

As I have already mentioned, the Babylonians measured the positions of the moon and planets with respect to one of the normal stars. In doing so, they specified both a direction and an angular distance, and we need to study the way they did both. Let us start with the way they specified the direction.

They used four directions: in front of, behind, above, and below. With regard to the two latter, Neugebauer [1975, pp. 546 - 547] writes: "Serious difficulties are caused by the expressions 'above' and 'below' a Normal Star, . . . If 'above' and 'below' referred to a sidereally fixed direction, the directions from a given star to any planet above or below it should always coincide." He then gives an example: On -197 October 29, Mars was recorded as being below Θ Leonis. The star was at longitude 132°.86 and latitude 9°.65, while the planet was at longitude 135°.4 and latitude 1°.8. The line connecting these two points meets the ecliptic at a longitude about 3° greater than does the orthogonal (to the ecliptic) drawn from the star.

Later, Aaboe [1980, p. 21] writes: "We are not quite sure precisely what 'above' and 'below' meant." He does not seem to be aware of my study [Newton, 1976, Section X.4] of this point, a study that of course was not available to Neugebauer in 1975. I am afraid that I do not understand why this point should be considered difficult.

In fact, if "above" and "below" are terms whose meanings are hard to understand, "in front of" and "behind" should also be difficult. However, they are not mentioned in this connection. To see what all these terms mean, let us return to the normal stars. These stars lie close to the ecliptic, and an astronomer who was well acquainted with them would automatically trace the plane of the ecliptic whenever he looked at the heavens and recognized some of them. Thus, when he wrote "in front of" or "behind", it seems natural to assume that he referred to the component of separation parallel to the ecliptic, with "in front of"

meaning to the west and "behind" meaning to the east.†
Similarly, it is natural to assume that "above" and "below"
refer to the component normal to the ecliptic.

In other words, if a Babylonian astronomer said that A
was 5° in front of B, he meant that the astronomical longitude of A was 5° less than (west of) that of B. If he wrote
that A was 5° above B, he meant that the astronomical latitude of A was 5° greater than that of B. Before I describe
the tests that establish the meanings of "above" and "below", let us turn to the units used to measure angles.

There are two angular units in the literature. The
names of the units are translated in various ways, so I
leave them untranslated. In transcriptions of the texts
into Latin letters, the units are rendered as "uban" (plural
ubanu) and "ammat" (plural ammatu). I will use these terms,
which will be abbreviated as u and am in tables.

The uban, which is the smaller unit, equals 1/12 of a
degree. It is often translated as digit, finger, or finger-
breadth. The ammat, which is the larger unit, is often
translated as the cubit or ell. The value of the ammat is
not uniform in Babylonian astronomy [Neugebauer, 1955, p.
39]. In some texts it equals 2° and in others it equals
2°.5, with the latter value tending to become more common in
the later literature. This means that we must establish the
value of the ammat separately for each text.

We can readily establish the value used in any observational text if we have a few measurements that use "above"
and "below", and at the same time determine if my meanings
of these terms are correct. The reason we can do this is
that the latitude of the moon or a planet is almost independent of the astronomical accelerations, so we can calculate
it with high confidence even for a period as remote as the
Babylonian period. We can certainly calculate the latitude
of a star, and thus we can calculate the difference in
latitude.

In the study of these terms that I mentioned above, I
was able to study texts from the four Babylonian years that
began in the spring of our years -567, -378, -273, and -231.
In all cases, the measured distance of moon or planet above
or below a star was directly proportional to the difference
in latitude, with "above" corresponding to positive differences. The ammat was clearly equal to 2° in the oldest
text, and it clearly equalled 2°.5 in the two latest years.
In the intermediate year -378, there was only one measurement available for test. On -378 November 13, the moon was
recorded as being 1 ammat below β Gemini, and the calculated
latitude of the moon minus that of the star is -2°.3. The
ammat probably equalled 2°.5 in this year also, but we cannot say so with high confidence.

---

†The Babylonian astronomers undoubtedly considered that the
heavens rotate from east to west, so if object A is in
front of object B, A is west of B.

The tests that establish the value of the ammat clearly confirm that "above" and "below" refer to differences in latitude, with the sign of the difference being just what one would expect, and with no implication about the difference in longitude.

3. The Babylonian Calendar

I gave most of what we need to know about the Babylonian calendar in Section VI.2. Here we need to consider briefly only the Babylonian method of designating the year, and the correspondence of the Babylonian years with ours. The Babylonian year began near the vernal equinox, and when I refer to a Babylonian year as equalling our year -378, for example, I mean the year that began in the spring of -378 and that continued into the spring of -377.

Until the year -310, the Babylonians identified a year by saying that it was, for example, the 5th year of the ruler Cambyses. From -310 onward, however, the years were numbered continuously, and the first day of the Babylonian year that began in -310 is called the Seleucid Era in modern writing about the Babylonian calendar.†

Translating a year of the Seleucid Era into our calendar is a trivial task for the historian. Matters are more difficult for earlier years. Once a single year is established in some reign, we can obviously extend the correspondence to all years of that reign. We cannot extend it to a different reign unless we have a Babylonian document that explicitly relates the two reigns.

In order to establish a correspondence for some year that has not been previously tied to our calendar, we need an event described and dated in a Babylonian document that can be identified and dated in our calendar. Here astronomical observations are particularly valuable. If we have observations of several different objects whose Babylonian dates can be connected, we can almost always establish a unique date for the text. I do not know the relative role that astronomical events and other events have played in establishing the Babylonian chronology.

At the present time, we can establish the year that corresponds to any Babylonian year since -625. Ptolemy gives a list of correspondences going back to the year -746, which he says was the first year of the king Nabonassar. For some time, Ptolemy's list was considered authoritative. We remember from Section VI.2 that Ptolemy gives "records" of eclipses allegedly observed in Babylon, that he gives the years in the Babylonian fashion and the complete dates in

---

†A Seleucid Era is still used in some religious calendars, with dates that may differ from both this date and from each other. Seleucus was the one of Alexander's generals who seized control of Babylonia and some other areas after the death of Alexander. See Appendix IV.

the Egyptian fashion, and that modern calculation shows that the eclipses took place in reasonable agreement with Ptolemy's records. The fact that the years Ptolemy gives agree exactly with his list of correspondences was considered definitive proof that the list is correct.

Now we know that Ptolemy fabricated the records in question, and this changes the situation. I want to review carefully what I wrote earlier on this matter [Newton, 1977, Section XIII.9], because I have been widely misquoted in book reviews and other commentaries on my work.

When Ptolemy wanted to use an "observation" of an eclipse made in, say, -720, he naturally attributed it to Babylon since he knew that no systematic Greek observations were being made that early. He first calculated the time and other circumstances in his own calendar, and since his value for the length of the month was rather accurate, his calculation of the time is reasonably accurate. After he had the date, he looked in his table of correspondences and assigned the Babylonian year from it but, as I explained in Section VI.2, he could not assign the Babylonian month and day.

In modern times, we calculate the time of an eclipse and find that it verifies Ptolemy's statements. Hence we conclude that his list of correspondences is accurate, and this conclusion would be valid if Ptolemy's records were valid. However, since Ptolemy made up his "observations", their agreement with modern calculation proves nothing about the validity of his calendrical correspondences.

It is quite possible that his list is accurate in spite of this, as I was careful to point out, but we cannot use Ptolemy as an independent authority for chronology. We can use his list only for those parts where we have independent confirmation. Further, we do have important independent confirmation. We have an extensive set of astronomical observations [Neugebauer and Weidner, 1915] that I have studied before [Newton, 1976, Chapter IV]. These observations are dated in the 37th year of Nebuchadnezzar (or Nebuchadrezzar) II, and they can be dated unambiguously to the year -567. This puts the first year of Nebuchadnezzar in -603, and this agrees with Ptolemy. Since his list is accurate in -603, it is reasonable to assume that it is substantially correct in later years, but we may not use it as a basic reference for earlier years. This does not say that it is necessarily incorrect for earlier years, but we must determine its accuracy by independent evidence; we may not use it as evidence by itself.

I gather that the reign of Nebuchadnezzar is tied firmly to that of his predecessor Nabopolassar, so we can carry the correspondence back to -625, still in agreement with Ptolemy. For earlier years, we are on shaky ground in correlating Babylonian chronology with our own.

Parker and Dubberstein [1956] give the dates in the Julian calendar of the first day of every Babylonian month from -625 to +75.† In doing this, they rely upon documentary sources, whenever possible, in saying which years have 13 rather than 12 months. However, they rely entirely upon calculation for determining the first day of the month. As they recognize, this introduces an uncertainty of a day, or of two days on rare occasions, in the beginning of a month. This uncertainty has little importance for most historical studies, but it can be important in astronomical work. We can use the dates of Parker and Dubberstein as the starting point in finding the dates of astronomical observations, but we must ultimately rely upon the observations themselves to furnish the exact dates.

4.   Some Observations from the Year 60 of the Seleucid Era

        Since making my earlier study of Babylonian observations, I have seen two more bodies of observations in the translated literature. One contains only three isolated observations but the other is rather extensive. I take up the extensive one in this section.

        This body of observations comes from an astronomical diary from the second half of the year 60 of the Seleucid era. An astronomical diary covers half of a Babylonian year and gives a variety of observations. The observations for each month form a separate section, which always starts by saying how many days were in the preceding month, and the interval between sunset and moonset on the first day of the month. Then, confining ourselves to the moon, the diary usually gives the position of the moon with respect to a star on several nights during the month, either in the evening or in the morning. In addition, near the full moon, it gives the interval from moonset to sunrise, from sunrise to moonset, from moonrise to sunset, and from sunset to moonrise, on the appropriate dates. It also gives the interval from moonrise to sunrise on the last morning the moon could be seen before the new moon. At this point, it often tells us about the weather and gives the prices of various staple foods.

        In its original form, the diary in question should have had a section for every month in the second half of the year 60 (autumn of -251 to spring of -250). However, the fragments that remain and that have been published in translation [Aaboe, 1980]‡ cover only about two and a half months. The sections for months VII and XII have been preserved

---

†Parker and Dubberstein do not say why they chose to begin their tables in -625, but I note with interest the coincidence with the conclusion I just reached. Their choice could not have been influenced by my conclusion, since their work is 21 years earlier.

‡The translation Aaboe presents is described as a preliminary translation made by A. Sachs. I do not believe that the definitive translation, when it appears, will alter our use of the data.

pretty well, but only about half of the section for month VIII is usable. The sections for the other months have not been discovered.

Before we can use the observations, as I have explained earlier, we must establish the exact relation between the Babylonian calendar and our calendar, and we must do this separately for each month. In addition, before we can use the measured positions of the moon, we must establish the value of the ammat used in the diary in question. It will greatly simplify the presentation of the data in tabular form if I state immediately that the calendar dates for all three months agree exactly with the tables of Parker and Dubberstein [1956], which are based upon calculations of the visibility of the new moon. Further, the diary says that month VII had 30 days, and this confirms the conclusion.† I do not believe it is necessary to give the details by which I verified the tables.

TABLE VII.1

THE VALUE OF THE AMMAT IN THE SELEUCID YEAR 60

Date		Star	$\Delta \beta$[b]		
Babylonian calendar[a]	Julian calendar[c]	used	Observed, ammatu	Calculated °	°/am
VII 17	-251 Oct 20	γ Gem	+2	+5.27	2.63
VIII 10	-251 Nov 11	β Ari	-6	-14.32	2.39
VIII 15	-251 Nov 17	α Gem	-4	-10.90	2.73
XII 3	-250 Mar 2	η Tau	-3	-8.03	2.68
XII 6	-250 Mar 5	γ Gem	+2	+5.81	2.91
XII 7	-250 Mar 6	α Gem	-3	-9.96	3.32
XII 9	-250 Mar 8	ε Leo	-3	-7.48	2.49
XII 10	-250 Mar 9	α Leo	+1	+2.61	2.61
XII 12	-250 Mar 11	β Vir	+1	+3.66	3.66
XII 18	-250 Mar 18	α Sco	+3 $1/2$	+6.74	1.93

[a]In the year 60 of the Seleucid era.

[b]The latitude of the moon minus the latitude of the star.

[c]When the reader compares the dates, he should remember that the Babylonian day began at sunset but the day in the Julian calendar begins at midnight.

---

†In addition, the diary closes by saying that month XII had 29 days, and by giving the interval from sunset to moonset on the first day of the next month. These data also agree with the calculated tables.

Matters are somewhat different with the value of the ammat. As I said in Section VII.2, the ammat was equal to 2° in a diary from the year −567, and it equalled 2°.5 in diaries from −273 and −231. Since the diary in question lies between the later two years, we expect the ammat to be 2°.5 in it also. However, Aaboe says specifically that the ammat equals 2° in this text, so we must look at the value in detail.

TABLE VII.2

LUNAR CONJUNCTIONS IN THE SELEUCID YEAR 60

Date	Hour[a]	Star	Longitude of star °	$\Delta\lambda$[b] As recorded	°
−251 Oct 5	15.573	θ Oph	230.109	−1.5am	−3.75
−251 Oct 14	15.387	η Psc	355.547	−1am	−2.50
−251 Oct 17	2.386	η Tau	28.719	?1am	−2.50
−251 Oct 18	2.399	α Tau	38.478	+2?.5am	+3.75
−251 Oct 25	2.500	θ Leo	132.105	−1.5am	−3.75
−251 Oct 28	2.544	α Vir	172.580	−1?am	----
−251 Nov 6	14.991	δ Cap	292.155	−0.5am	−1.25
−251 Nov 10	14.939	η Psc	355.546	−3am	−7.50
−251 Nov 13	14.905	η Tau	28.720	+2am	+5.00
−251 Nov 16	2.844	μ Gem	63.989	+8u	+0.67
−250 Feb 28	15.729	β Ari	2.691	−2am	−5.00
−250 Mar 3	15.764	α Tau	38.475	+2am	+5.00
−250 Mar 5	15.788	γ Gem	67.804	+0.5am	+1.25
−250 Mar 6	15.799	α Gem	79.043	−0.5am	+1.25
−250 Mar 9	15.834	α Leo	118.701	+0.83am	+2.08
−250 Mar 10	15.844	ε Leo	109.416	−0.83am	−2.08
−250 Mar 14	2.615	α Vir	172.596	+1.5am	+3.75
−250 Mar 16	2.571	β Lib	198.125	−0.83am	−2.08
−250 Mar 17	2.548	β Sco	211.916	−1.83am	−4.58
−250 Mar 19	2.503	θ Oph	230.123	+1am	+2.50
−250 Mar 22	2.433	β Cap	272.760	−0.5am	−1.25
−250 Mar 24	2.387	δ Cap	292.160	+3.0am	+7.50

[a]In Greenwich mean time.

[b]Longitude of the moon minus the longitude of the star.

The diary says that the moon was "above" or "below" a particular star on ten different dates.† The data for these dates are summarized in Table VII.1. The first column gives the Babylonian date; since all the dates are in the Babylonian year 60, only the Babylonian month and day are given. The second column gives the date in the Julian calendar at the time of the observation. In comparing the dates, the reader should remember that the date in the Babylonian calendar begins at a sunset but that the date in the Julian calendar begins at the following midnight. The third column gives the star to which the moon was referred.

The next two columns give the latitude of the moon minus the latitude of the reference star, as recorded and as calculated. In the column for the recorded value, the unit is the ammat. When the record says that the moon was above the star, this is represented by the + sign. The calculated value is based upon using −28 for the acceleration of the moon (with respect to ephemeris time) and −10 for the value of y (the acceleration of the earth's spin in parts in 10^9 per century). I meant to use −20, but by mistake I copied from the computer listing for −10. Since the latitude depends but little upon the acceleration used, it did not seem worthwhile to remake the table after I discovered this mistake.

The last column is the ratio of the two preceding ones. That is, it gives the number of degrees per ammat indicated by each observation. It is clear from inspection that the value is 2°.5 rather than 2°, in agreement with my earlier finding for the years −273 and −231 and in disagreement with Aaboe.

The date for which the ratio is most discordant with 2°.5 is XII 12, but the recorded difference for this date is only 1 ammat. The dates for which the observations are most discordant with modern results are XII 7 and XII 18, for which the discrepancies exceed 2°. It is not unusual with records this old to find recording errors one time in five, and these values may in fact be recording errors. If we omit these values, it does not affect our conclusion about the value of the ammat; it only improves our feeling about the accuracy of the observations.

Now that we have established both the dating of the observations and the size of the angular unit used, we may turn to the records of the lunar conjunctions that give a difference in longitude between the moon and a star. These records are summarized in Table VII.2. I have omitted a few conjunctions of the moon with a planet. The slight extra strength that these conjunctions would give to our results does not justify the extra complexity that arises from using a planet instead of a star.

---

†Incidentally, saying that the moon was above or below a star does not imply that the observer thought the difference in longitude was zero. In fact, he stated a difference in both latitude and longitude on three dates out of the ten.

In Table VII.2 the first column gives the dates in the Julian calendar. The reader should have no trouble in converting them to the Babylonian calendar with the aid of Table VII.1 if he should wish to do so, provided that he pays due regard to the hour of the observation. The hour itself is given in the second column. Sometimes the records say that the observation was made at the beginning of the night or at the end of the night, and sometimes the records say only in the evening or morning. I think the same general circumstance is meant in all cases, namely a time after sunset when the reference stars have become visible, or a time before sunrise when they are still visible. In my earlier study [Newton, 1976, Section IV.7], I decided to take the hour to be 45 minutes after sunset or 45 minutes before sunrise,† whichever is appropriate, and I do the same here. Since the observations are evenly balanced between morning and evening, a reasonable error in this assumption has little effect upon the results.

The third column gives the modern name of the star used and the fourth column gives the calculated longitude of the star with respect to the true equinox of date. It is probably an unnecessary refinement, but I included the annual aberration in calculating the longitude. The fifth column gives the recorded difference in longitude between the moon and the star, except that I have replaced "in front of" or "west of" by a minus sign and "behind" or "east of" by a plus sign. I have also written fractions in decimal form instead of the form the recorder used. The last column gives the recorded difference converted into degrees. Four entries need discussion.

The entry for -251 October 17 is damaged and the translator could not make out whether the moon was in front of or behind the star. To anticipate later results, calculation shows that the moon was west of the star, so I have felt it safe to supply a minus sign in the last column.

Likewise, the entry for -251 October 18 is damaged and the translator could not make out the integer part of the value. He thought it looked more like 2 than any other integer, so he stated this opinion by writing "2?" for the integer part. Calculation shows that the interval was rather close to +1.5 ammatu and I have felt safe in taking this to be the reading. This is equal to 3°.75.

The same thing happens with the entry for -251 October 28, except that the translator thought the integer was probably 1. Here, calculation shows that the separation was much larger in magnitude than 1 ammat, and we cannot safely restore the uncertain integer. I have indicated this by leaving a blank in the last column.

---

†I define sunrise and sunset with respect to the upper limb of the sun.

Finally, the record for -250 March 6, after saying that the moon was 3 ammatu below α Gemini, says that it was also ½ ammat back to the west. If "back" means the same thing as the term I have rendered as "behind" in other records, this is a contradiction in terms. "Behind" means that the moon was behind in the diurnal motion and hence that it was east of the star, not west.

I believe that "back" as used here is the same as a term that occurred in my earlier study of Babylonian observations, and I will defer the study of this term to Appendix V. Here I say only that the moon, as is clearly shown by calculation, was east of α Gemini rather than west. I have indicated this by changing the sign of the separation between the last two columns in Table VII.2. However, we may not take it as a general rule that "back" implies "east" in this context. See the appendix.

TABLE VII.3

LUNAR SYNODIC PHENOMENA IN THE SELEUCID YEAR 60

Date	Later event[a]	Earlier event[a]	Measured time interval h
-251 Oct 3	MS	SS	0.867
-251 Oct 15	SS	MR	0.278
-251 Oct 16	SR	MS	0.256
	MR	SS	0.156
-251 Oct 17	MS	SR	0.667
-251 Oct 30	SR	MR	1.333
-251 Nov 14	SR	MS	0.667
	SS	MR	0.200
-251 Nov 15	MS	SR	0.222
-250 Feb 28	MS	SS	
-250 Mar 12	SR	MS	0.533
-250 Mar 13	SS	MR	0.467
-250 Mar 14	MR	SS	0.422
-250 Mar 26	SR	MR	0.867
-250 Mar 29	MS	SS	1.389

[a]MR = moonrise, MS = moonset, SR = sunrise, SS = sunset.

In addition to recording the position of the moon at frequent intervals during the month, the Babylonian astronomers also recorded certain synodic phenomena, that is, phenomena which relate the position of the moon to that of the sun. Aside from eclipses, which do not occur in the surviving part of the diary from the Seleucid year 60, six synodic phenomena were observed every month, weather permitting; these are the phenomena listed near the beginning of this section. These phenomena are often called the "lunar sixes" in modern writing [Sachs, 1948].

The Babylonian astronomers also recorded synodic phenomena that relate a planet to the sun, but we do not deal with planetary data in this work.

The lunar synodic phenomena in the diary from the year 60 of the Seleucid era are summarized in Table VII.3. In the table, the first column gives the date of the observation in the Julian calendar. The next two columns specify the two events involved, with the later event in the second column and the earlier event in the third column. The last column gives the measured time interval between the two events.† Thus, for example, on -251 October 3, the time interval from sunset to moonset was 0.867 hours.

All of the lunar sixes survive for month VII, which is the first month in the diary. The diary makes the interesting remark that sunset to moonset on the first day of month VIII was expected to be 18;20 degrees, ". . . but because of clouds I did not watch." It also says that month VII had 30 days. The rule was that a new month began if the moon was visible at the sunset that closed the 29th day. If the moon was not visible at this time, the new month did not begin yet, whether the skies were cloudy or not. However, a new month began at the following sunset whether the moon was seen or not. Thus no month had more than 30 days.

In this case, we know that the new month began after 30 days, even though the moon was not seen at the relevant sunset. It was not seen on this occasion because of clouds. Also, it was not seen on the preceding evening, because the month did not begin then. However, we do not know whether the failure to see it on the preceding evening was caused by clouds or not.

The record for month XII also says that from sunrise to moonset on the 13th day was expected to be 1;10 degrees, " . . . but because of clouds I did not watch." The other five events that month were measured, but the time interval from sunset to moonset on the first day cannot be read. The diary also gives the interval from sunset to moonset on the first day of the first month of the new year.

5.   Three Observations from the Year 23 of Artaxerxes I

van der Waerden [1974] gives a few extracts from an astronomical diary for the first half of the 23rd year of Artaxerxes I. The only data that are useful for our purposes are: Nisannu, the new moon was visible for 15 degrees. Aiaru, the new moon was visible for 24 degrees. Simanu, the new moon was visible for 16 degrees. Nisannu

---

†The unit used in recording the time interval is often called the time degree in the modern literature. It equals 4 minutes. I have changed the recorded intervals to hours.

was the name of the first month of the year, and it started
sometime in the spring. Aiaru was the second month and
Simanu was the third month.

We know from the tables of Parker and Dubberstein that
the 23rd year of the king Artaxerxes I began in April of
-441. With this as our starting point, we can readily establish the dates of these new moons. The tables prove to be
correct for all three months. When the record says, for
example, that the new moon was visible for 24 degrees, it

TABLE VII.4

THREE NEW MOONS FROM THE 23RD YEAR OF ARTAXERXES I

| Date | | Interval, |
Babylonian calendar	Julian calendar	sunset to moonset h
I 1	-441 Apr 10	1.000
II 1	-441 May 10	1.600
III 1	-441 Jun 8	1.067

---

means that the interval from sunset to moonset was 24 x 4 =
96 minutes, or $1^h.6$.

The observations are summarized in Table VII.4. The
first column gives the Babylonian month and day (which is
the first day of the month for all observations); the year
is the same for all three observations. The second column
gives the date in the Julian calendar, and the third column
gives the measured interval from sunset to moonset, converted to hours.

6.   Two Timed Solar Eclipses

Three modern sources [Kugler, 1924, p. 385; Fotheringham, 1935; Muller and Stephenson, 1975] have discussed a
record of the solar eclipse of -321 September 26.† The
eclipse was in the second year of Philip, on the 28th day of
month VI. By the tables of Parker and Dubberstein [1956],
this is equivalent to -321 September 26. This is the date
of an annular eclipse that was barely visible in Babylonia
near sunset [Oppolzer, 1887]. The modern sources disagree
about two points in the record, which is tablet BM 34093.

---

†Muller [1975, pp. 8.26ff] also discusses this record, but
his discussion is drawn from his earlier joint work with
Stephenson.

Kugler says that the eclipse began 4 time degrees before sunset. Fotheringham quotes Kugler on this point, and says that "Dr. Langdon" has kindly verified the reading.†
Muller and Stephenson say that the eclipse began 3 time degrees before sunset. They note the discrepancy in time, and they say that Professor P. J. Huber examined the original tablet and confirmed the reading of 3 time degrees.

Since we have an equal weight of the authorities up to this point, I asked Dr. Walker of the British Museum‡ to examine the tablet. His reply makes two main points. First, Professor A. Sachs of Brown University joined tablet BM 34093 to tablet BM 35758 in 1954, so one should now refer to BM 34093 + 35758. Second, the reading is 3, not 4, time degrees before sunset. Thus, according to the record, the eclipse began 12 minutes before sunset.

Fotheringham says that the record "is believed to come from Sippara" while Muller and Stephenson take it for granted that it comes from Babylon. Neugebauer [1955, volume 1, p. 5] says that the collection of texts from which the record comes was originally assigned to Sippar (Sippara) on the basis of a reading by Kugler. Now, Neugebauer writes, Kugler's reading "turns out to be a mistake, and we are therefore left with no direct evidence for astronomical texts from Sippar." I therefore take Babylon to be the place; the distance between Babylon and Sippar is not important for our purposes.

Since the measured time interval is only 12 minutes, this is the most precise measurement of the time of a solar eclipse that we have before the Islamic period. Muller [1975, p. 8.27] notes this, and then writes that the sensitivity of the time to the variations in the accelerations is too weak for the record to be useful, in spite of its high precision.

I do not understand this remark. When the path of a solar eclipse has a portion that lies in an east-west direction, and when the observer is near this portion of the path, the magnitude he observes is insensitive to the accelerations. The times, however, are always sensitive to them. The sensitivity depends strongly upon the date of the eclipse, but it is nearly the same for all eclipses of about the same date,‡ regardless of the geometry.

---

†This is probably S. Langdon, who wrote a book jointly with Fotheringham in an attempt to date some early observations of Venus. See Newton [1976, Section IV.1] for a discussion of their work and of other attempts to date these records.

‡This is C. B. F. Walker, Assistant Keeper. I thank Dr. Walker for his kind assistance in this matter.

‡This is certainly true at low latitudes, where Babylon was. We can have peculiar effects at high latitudes, and it may be that the statement is not always true in the polar regions; I have not tried to investigate the point.

Muller and Stephenson [1975, p. 482] have published two records which they relate to the eclipse of -135 April 15. Again the discussion of these records by Muller [1975, pp. 8.30 - 8.31] is derived from his joint work with Stephenson. Muller and Stephenson also note an earlier publication of a drawing based upon one of the cuneiform tablets.

After identifying the year as 175 of the Seleucid era (-136/-135), one record reads: "Daytime of the 28th the north wind blew. Daytime of the 29th, 24 uš† after sunrise, a solar eclipse beginning on the southwest side . . . 35 uš for obscuration and clearing up . . ." The record also says that Mercury, Venus, Jupiter, and Mars were visible during the eclipse, which implies a large but not necessarily total eclipse. Professor A. J. Sachs provided this translation to Muller and Stephenson in a private communication. At least in the part that has been published, this record does not state the month. However, during the year that is stated, there is only one possible eclipse, that of -135 April 15. According to the tables of Parker and Dubberstein, this date came in the month called second Addaru.‡ This eclipse was certainly large in Babylon and it may have been total.

The other record gives the month as the second Addaru but the surviving part does not give the year. The part relating to the eclipse is given as: "On the 29th day there was a solar eclipse beginning on the south-west side. After 18 us . . .it became complete such that there was complete night at 24 uš after sunrise." Professor Huber provided this translation to Muller and Stephenson in a private communication, in which he also identified the year as -135 by means of references to Venus and Mars.

Muller and Stephenson say that the 29th day of the second Addaru in -135 is equivalent to -135 April 15, according to the tables of Parker and Dubberstein [1956]. This is not correct. According to the tables, the first day of the second Addaru was -135 March 19, so that the 29th day was -135 April 16, not April 15. If the record is correct, the first day of the month was March 18 rather than March 19. Further, Parker and Dubberstein give the first day of the preceding month as February 17, which is 29 days before March 18. We remember that a month had 29 days only if the new moon was actually observed at sunset on the 29th day.

----

†This is the unit called the "time degree", which equals 4 minutes.

‡The Babylonian months were lunar but the years were solar on the average. To high accuracy, 235 (synodic) months equal 19 years, and 19 x 12 = 228. Hence 7 years out of every 19 must have 13 lunar months in order to keep the calendar year in approximate synchronization with the solar year. At this stage in history, the extra month was called "second Addaru" six of the seven times that an extra month was needed, and it was called "second Ululu" the other time.

Hence, if February 17 is correct, the new moon was actually observed at the sunset that preceded March 18. (That is, the new moon was observed at sunset on -135 March 17.) If February 17 is not correct, second Addaru had 30 days, and the new moon was observed at the sunset that preceded February 16. (That is, it was observed at sunset on February 15.) At the moment, we cannot settle between the two possibilities.

Muller and Stephenson also say that the two records between them give a highly reliable description of a total eclipse that is a "testimony of (sic) the skill of the Babylonian astronomers." I am afraid I cannot bear witness to this testimony. The records seem contradictory to me. The first record seems to say that the eclipse began 96 minutes (24 uš) after sunrise and that it lasted 140 minutes. The second record seems to say that the eclipse began 72 minutes after sunrise and that it was total 24 minutes later, at 96 minutes after sunrise. This is impossible.†

In my earlier study of this record (volume 1, pp. 204ff), where I took it for granted that both records refer to the same eclipse, I developed a contrived interpretation that reconciled the times given in the two records. Now I am inclined to doubt that the records are of the same eclipse. The fact that both eclipses came on the 29th of the month is not significant in a lunar calendar, since a solar eclipse necessarily comes close to the end of the month if the month ends with the visibility of the new moon. It is also not highly significant that both eclipses came in the second Addaru, since almost one year in three had this intercalary month. Finally, we may not assume that the second record refers to a total eclipse, in spite of its use of "complete night". In the earlier work just mentioned (p. 96), I studied the quantity 1 - m (in which m is the greatest magnitude of an eclipse) for records which assert specifically that an eclipse was total. The standard deviation of 1 - m for such records was 0.071 for Chinese records and 0.034 for records from other provenances. Thus an eclipse reported as total was often far from total.

Since the references to Venus and Mars have not been published, we cannot judge the strength of the evidence that indicates -135 as the year. Pending the publication of this evidence, I will assume that the second record does not belong to the year -135 and I will use only the first record. I will assume that the eclipse either began 96 minutes after sunrise or that it became total then. If the calculations cannot settle unambiguously which assumption is correct, I will not use the time given in the first record.

---

†Since part of the text seems to be missing, we cannot say with certainty that "18 uš" refers to the beginning of the eclipse. However, there cannot be any significant stage of an eclipse that occurs 24 minutes before the beginning of totality.

However, I will use the duration stated if it turns out to
be sensitive to the accelerations.†

7.   Some Eclipse Records That Have Been Discussed Before

    In my earlier work [Newton, 1976, Table IV.3 and Sec-
tion X.2], I discussed several other records that refer to
eclipses, but I did not use any of them in estimating the
astronomical accelerations. I did use some of them in establishing the dates on which certain Babylonian months
began. Here I will make further use of some of the records,
and more discussion is needed.

    Kugler [1907, pp. 70 - 71] presents an astronomical di-
ary from the 7th year of Cambyses (-522/-521) that describes
two lunar eclipses. One came on the night of the 14th day
of month IV. This is -522 July 16 according to the tables
of Parker and Dubberstein.‡ The eclipse was 3 1/3 hours
after sunset and the entire course was visible. The northern half was eclipsed. I take this to be 6 digits of the
area, which from Figure A.II.2 equals 0.47 of the diameter.

    The other eclipse came on the 14th day of month X, but
it was after midnight and this equals -521 January 10 according to the tables. It was 5 hours "of the night toward
morning", which I take to mean 5 hours before sunrise. The
entire eclipse was visible, and it stretched over the southern and northern halves of the moon. Since it was total, we
cannot make any use of the magnitude.

    I will use both of these eclipses on the assumption
that the times were of the beginning. It is interesting
that Ptolemy chose to use the eclipse of -522 July 16 in his
work, but the circumstances he gave for it do not agree with
those in the cuneiform text. This caused much trouble in
the literature until it was realized that Ptolemy had fabricated the "record" published in his work [Newton, 1977, Section XIII.9]. See Section VI.2 above.

    Schaumberger [1933] has published a text that refers to
a lunar eclipse on -248 April 18, a solar eclipse on -248
May 4, a lunar eclipse on -248 October 13, and a solar
eclipse on -248 October 27. The record says that the two
lunar eclipses did not happen, and that the solar eclipse on

---

†It will turn out in Section VIII.5 that the eclipse began
at 96 minutes after sunrise. It seems to me that this is
incompatible with the second record, if the translation is
correct. Since we now know the beginning time, I will not
use the duration of the eclipse, since it must be much less
sensitive than the beginning.

‡The tabular date is actually -522 July 17, but we must
subtract one day because the eclipse started before
midnight.

-248 October 27 began 4^h 28^m after sunset.† It is clear
that these parts of the text came from calculation and not
observation. The reference to the other eclipse is harder
to judge.

The surviving part of the relevant text reads: ". . .
on the 28th an eclipse of the sun . . . 6 hours after sunrise." The month has already been identified as the first
(Nisannu), so the date is equivalent to -248 May 4 according
to the tables of Parker and Dubberstein. There was an
eclipse on this date that was visible in Babylon, weather
permitting. However, this record has the same form as the
records that are clearly calculated, with nothing to indicate observation. Hence, the safe procedure is to assume
that this record is also calculated.

Schaumberger has also published a text which refers to
a solar eclipse on -239 October 18 and a lunar eclipse on
-239 November 3, and there were eclipses on these dates.
The record of the solar eclipse says that it was after sunset, so this record was clearly calculated. The record of
the lunar eclipse says that it began 3 uš (12 minutes) before sunrise on the easterly side of the moon's disc. Now,
a Babylonian astronomer of this period could have known that
a lunar eclipse always begins on the east limb of the moon,
but this statement does not occur in either of the records
of lunar eclipses from -248, nor does it occur in any other
calculated record of a lunar eclipse from Babylon that I can
think of. Hence, I tentatively take this as evidence that
the eclipse of -239 November 3 was observed as stated. I
will return to this record in Section VIII.4.

Epping and Strassmaier [1892] have published a trans-
cription and translation of an astronomical diary from the
Seleucid year 79 (-232/-231) that refers to the solar
eclipse of -232 November 30. It should have been at 44 time
degrees (2^h 56^m) after sunrise, but it was not seen. Actually there was a penumbral solar eclipse whose center time
came at about the time stated, but which was probably not
visible as far south as Babylon. The diary also says that
there was a lunar eclipse on -232 December 14 at 4^h 56^m
after sunrise. There was a lunar eclipse on this date whose
center time was about noon in Babylon. It is clear that
neither of these records constitutes an observation.

Epping and Strassmaier [1891a] have published a tran-
scription and translation of another text from the Seleucid
year 100 that refers to three eclipses. First is a lunar
eclipse on the 13th day of the first month about 80 minutes
before sunrise; the moon was eclipsed on the south side.
This date is equivalent to -211 April 30, and I think it is
safe to take this as an observation.

----

†These statements are essentially correct. There were no
  lunar eclipses in -248. There was an annular solar eclipse
  on -248 October 27 that occurred during the nighttime at
  Babylon. I have not checked the hour of this eclipse.

Next the record says there should have been a solar
eclipse on the 28th of the same month and hence on -211 May
15, 140 minutes before sunset, but it was not seen. Finally, the text calls for a lunar eclipse on the 15th day of
month VII, and hence on -211 October 24 (it came before
midnight). The next part is hard to read, but the translators think it says "northeast side"; this reading indicates observation if it is correct. The difficulty concerns
the time, which the text says was 28 time degrees (112 minutes) after sunset. According to the calculations made by
Epping and Strassmaier, this was at $19^h\ 15^m$ local mean time.
According to Oppolzer [1887], the eclipse did not begin until $18^h\ 50^m$ Greenwich time, which is about $21^h\ 45^m$ local
time. This is far too big a discrepancy to account for
either by the approximations in Oppolzer's tables or by an
error in measurement.

Epping and Strassmaier incline toward the opinion,
which seems reasonable to me, that the eclipse was observed,
but that an error was made in writing the time. Even if
this is so, though, there is nothing from the surviving
record we can use.

A third text published by Epping and Strassmaier
[1891b] also refers to two solar and two lunar eclipses in
the Seleucid year 207 (-104/-103). First, a solar eclipse
on the equivalent of -104 July 29 came 4 hours after sunset,
and indeed the eclipse of that date was visible only in the
South Pacific. Second, a lunar eclipse on -104 August 13
came at 56 minutes before sunrise and hence at about $4^h\ 15^m$.
This agrees well with Oppolzer, who shows the eclipse of
that date beginning at almost exactly 4^h, Babylon time, on
that date. Third, there was a lunar eclipse on -103 January
8 at 100 minutes before sunset which was therefore not visible. In fact, there was no eclipse on that date. Finally,
there was a solar eclipse on -103 January 22 that did not
occur; the path of the eclipse of that date began in mid-
Pacific, crossed South America, and ended up near the
Equator in the Gulf of Guinea.

Here, we have a situation like that in -248. Of four
eclipses mentioned in a uniform context, three could not
have been observed. The other one could have been, but
there is no remark to indicate observation. Hence, we
should not use any of the text from this year.

The last text we need to consider comes from an earlier
time. Kugler [1914, pp. 233 - 242] published a text that
includes a large amount of astronomical information, including statements about two eclipses. The text, as Kugler
shows, is from the 40th year of Artaxerxes I (-424/-423).
One line in the text says that the lunar eclipse of -424
October 9 began 40 minutes after sunset. The next line says
that there was a solar eclipse on -424 October 23, but says
nothing about the hour. We can find credible values of the
astronomical accelerations that will make either of the two
eclipses be visible in Babylon, but there is no set of

accelerations that lets both of them be visible,† and Kugler
himself establishes this point.

These lines occur in a text that has all the appearance
of a calculated ephemeris, both in the arrangement of its
material and the absence of any remarks indicating any observation at all. In particular, there is no remark re-
garding the side of the moon where the lunar eclipse began.
Thus there is no reason to assume that the lunar eclipse was
observed.

TABLE VII.5

BABYLONIAN RECORDS OF ECLIPSES[a]

Date	Type	Observed quantity	Observed value
-522 Jul 16	Lunar	Beginning Magnitude	3^h 20^m after sunset 0.47
-521 Jan 10	Lunar	Beginning	5^h before sunrise
-321 Sep 26	Solar	Beginning	12^m before sunset
-239 Nov 3	Lunar	Beginning	12^m before sunrise
-211 Apr 30	Lunar	Beginning	80^m before sunrise
-135 Apr 15	Solar	b Duration	96^m after sunrise 140^m

[a]The place of observation is taken to be Babylon for all
these observations.

[b]We cannot tell from the text whether this is the beginning
of the eclipse or the beginning of totality, but it is
probably the beginning of the eclipse.

Unfortunately, Fotheringham [1935] chose to ignore
Kugler's analysis of the text and took the statement about
the lunar eclipse to be an observation. He used it in his
studies of the secular accelerations, and other writers have
taken over his results. Thus, the calculations of a Babylonian astronomer about the year -424 ended up as part of the
data base used by Spencer Jones [1939] in the famous paper
that led to the concept of ephemeris time. Luckily, as it
happens, the use of this eclipse is equivalent to reasoning
in a circle, which merely seems to confirm a conclusion that
was already established. The conclusion that was already
established [Newcomb, 1875 and earlier writers] was based

†I have given the details elsewhere. See Newton [1976,
Section IV.9].

upon valid data and thus Fotheringham's reasoning in a circle did no essential harm to astronomy, only to history.

Table VII.5 summarizes all the Babylonian records of eclipses that have been discussed in this chapter and that seem to be the results of observation. The place will be taken as Babylon for all these eclipses. The records in Table VII.5 will be analyzed in Sections VIII.4 and VIII.5.

CHAPTER VIII

ANALYSIS OF THE OBSERVATIONS BEFORE THE ISLAMIC PERIOD

1. General Plan of the Rest of this Volume

Quantitative lunar observations between the Islamic and
the modern periods are relatively sparse, and I analyzed
them as we went along in Chapters II through IV. With only
a few exceptions, though, I have not analyzed the data presented in Chapters V through VII, and I now take up the analysis of the data presented in these chapters. This analysis
will be in chronological order, which is the opposite of the
order in which the data were presented.

I used a large body of Babylonian observations in an
earlier work [Newton, 1976, Chapters IV and X], and I use
them again here, with the adoption of the DE102 ephemeris.
Since I make little change in the interpretation of the
records, I put the analysis of the records into Appendix
V. However, I have completely rediscussed the Babylonian
records of eclipses. I will analyze these eclipse records
in this chapter, along with the records from the year 23 of
Artaxerxes I and the year 60 of the Seleucid era. In this
chapter, I will also analyze the Chinese record of the solar
eclipse of -88 September 29 and the few Greek observations
we have. Then I will analyze the Islamic records in Chapter
IX.

After I have completed this analysis, I will collect
all the equations of condition that have been derived, put
them into chronological order, and solve them within time
intervals that are as short as the data will allow. This
will give us the time history of the acceleration parameter
y. We must then see how the time history derived here compares with that of Newton [1979, Section XIV.1] and see what
physical conclusions we can draw.

In all this analysis, I use $\dot{n}_M = -28$. The reader, however, may wish to know what the results for y would be if
some other value had been used. In order to do this, he
needs to know the partial derivatives of the observed quantities with respect to $\dot{n}_M$ as well as y. For some types of
observation, such as the time of a lunar conjunction with a
star, the partial derivatives are fairly obvious, and I will
not always present them. For all observations for which the
partial derivative with respect to $\dot{n}_M$ is not obvious, I will
tabulate the desired derivatives.

2. The Synodic Lunar Observations from Babylon that Have Not Been Used Before

By a synodic lunar observation, I mean one that, at
least implicitly, gives the position of the moon relative to
the sun. Of course, an observation of an eclipse relates
the moon to the sun, but the analysis of eclipses poses

particular problems, and I will put eclipses into separate
sections. In this section, I will analyze the synodic
observations that give the time interval between either the
rising or the setting of the sun and moon.

TABLE VIII.1

SOME SYNODIC LUNAR OBSERVATIONS

Date	Nature of interval	Δ_m[a] (h)	Δ_c[b] (h)	$\partial\Delta_c/\partial\underline{y}$	$\partial\Delta_c/\partial\underline{\dot{n}}_M$
−441 Apr 10	MS − SS	1.000	0.852	−0.0088	+0.0066
−441 May 10	MS − SS	1.600	1.393	−0.0094	+0.0064
−441 Jun 8	MS − SS	1.067	0.921	−0.0092	+0.0060
−251 Oct 3	MS − SS	0.867	0.898	−0.0038	+0.0032
−251 Oct 15	SS − MR	0.278	0.258	+0.0044	−0.0032
−251 Oct 16	SR − MS	0.256	0.396	+0.0084	−0.0056
−251 Oct 16	MR − SS	0.156	0.331	−0.0046	+0.0032
−251 Oct 17	MS − SR	0.667	0.697	−0.0082	+0.0056
−251 Oct 30[c]	SR − MR	1.333	1.811	+0.0070	−0.0056
−251 Nov 14	SR − MS	0.667	0.951	+0.0080	−0.0050
−251 Nov 14	SS − MR	0.200	0.135	+0.0046	−0.0038
−251 Nov 15	MS − SR	0.222	0.109	−0.0080	+0.0056
−250 Mar 12	SR − MS	0.533	0.517	+0.0042	−0.0034
−250 Mar 13	SS − MR	0.467	0.364	+0.0068	−0.0056
−250 Mar 14	MR − SS	0.422	0.534	−0.0068	+0.0056
−250 Mar 26	SR − MR	0.867	1.060	+0.0056	−0.0034
−250 Mar 29	MS − SS	1.389	1.328	−0.0088	+0.0056

[a]The measured time interval.
[b]The interval calculated using $\underline{y}$ = −20 and $\underline{\dot{n}}_M$ = −28.
[c]This observation will not be used in the analysis.

Table VII.4 gives three synodic observations from the
23rd year of Artaxerxes I (−441), and Table VII.3 gives
fourteen observations from the Seleucid year 60 (−251/−250).
Table VIII.1 summarizes the analysis of these data. In the
table, the first column gives the date of the observation in
the Julian calendar. The second column gives the nature of
the time interval that was measured. For example, if we see
"MR − SS" in this column, it means that the Babylonian
astronomer measured the time of moonrise minus the time of
sunset. With this example given, I believe that the other
entries in this column are obvious.

The third column gives the interval, in hours, that the
astronomer measured, and the fourth column gives the interval

that we calculate. In the calculation, we use the parametric values $\underline{n}_M = -28$ and $\underline{y} = -20$. The last two columns give the partial derivatives of the calculated interval Δ_c with respect to $\underline{n}_M$ and $\underline{y}$.

We see immediately upon scanning Table VIII.1 that all of the differences between the measured interval Δ_m and the calculated interval Δ_c look reasonable except for -251 October 30. Here the difference approaches half an hour while the other differences are around 15 minutes or less. I assume that an error was made in recording the length of the interval for -251 October 30, and I omit this measurement from the analysis. This leaves 16 values for use.

We do not know what devices the Babylonian astronomers used in measuring the time, but they were probably rather primitive -- perhaps sand glasses or water clocks. Thus it is plausible that the error in measuring a time interval should increase with the length of the interval. I explore this question in Appendix V with the aid of other data, and can find no dependence of the error upon the time interval. The data in Table VIII.1 lead to the same result. Hence I give the same weight to all the observations.

In the analysis of Table VIII.1, let $\delta_y = \underline{y} + 20$. That is, δy is the difference between $\underline{y}$ and the standard value -20 that we used in preparing the table. If we use all the observations except that of -251 October 30, we get the following equation of condition:

$$0.0008\ 1128\delta y = -0.007\ 476. \qquad (VIII.1)$$

This immediately leads to $\underline{y} = -29.2$, at the approximate epoch -300. When we use this value of $\underline{y}$ and calculate the residuals of the measured values, we get the standard deviation of the residuals to be 0.124 hours. When we combine this with the coefficient of $\delta \underline{y}$ in Equation VIII.1, we get 4.4 for the standard deviation of the inferred value of $\underline{y}$.

In other words, the value of $\underline{y}$ from using all of Table VIII.1 (except -251 October 30) is

$$\underline{y} = -29.2 \pm 4.4 \text{ parts in } 10^9 \text{per century.} \qquad (VIII.2)$$

However, when I come to trace the time history of $\underline{y}$, I will put the observations from the year -441 into a different "time bin" from those of the year -251/-250. For the year -441, we have

$$\underline{y} = -38.3 \pm 7.8. \qquad (VIII.3)$$

For the year -251/-250, we have

$$\underline{y} = -25.1 \pm 5.2. \qquad (VIII.4)$$

It is interesting to look at the relative accuracy with which the time intervals were measured. We have already seen that the standard deviation of the measured intervals

is 0.124 hours. The average length of the intervals (that
is, of their absolute values) is about 0.666 hours. Thus
the relative accuracy was about 0.18, or about 18 per cent.

TABLE VIII.2

ANALYSIS OF LUNAR CONJUNCTIONS FROM
THE SELEUCID YEAR 60

| Date | Hour[a] | Longitude of the moon | |
		Measured °	Calculated[b] °
-251 Oct 5	15.573	226.359	226.554
-251 Oct 14	15.387	353.047	353.077
-251 Oct 17	2.386	26.219	27.357
-251 Oct 18	2.399	42.228	41.463
-251 Oct 25	2.500	128.355	130.525
-251 Nov 6	14.991	290.905	290.788
-251 Nov 10	14.939	348.046	347.946
-251 Nov 13	14.905	33.720	30.705
-251 Nov 16	2.844	64.656	63.178
-250 Feb 28	15.729	357.691	358.425
-250 Mar 3	15.764	43.475	42.338
-250 Mar 5	15.788	69.054	69.368
-250 Mar 6	15.799	80.293	82.264
-250 Mar 9	15.834	120.784	119.320
-250 Mar 10	15.844	130.024	131.359
-250 Mar 14[c]	2.615	176.346	171.187
-250 Mar 16	2.571	196.045	195.283
-250 Mar 17	2.548	207.333	207.501
-250 Mar 19	2.503	232.623	232.432
-250 Mar 22	2.433	271.510	271.767
-250 Mar 24	2.387	299.660	299.845

[a]In Greenwich mean time.
[b]Calculated with $\dot{n}_M = -28$ and $\underline{y} = -20$.
[c]The observation of this date is omitted from the analysis.

3. Measurements of Lunar Longitude from the Seleucid Year 60

Table VII.2 lists 22 Babylonian measurements of the
lunar longitude that have survived from the year 60 of the
Seleucid era; the measurements were made by comparing the
lunar position with that of a star. In the measurement of
-251 October 28, the text is ambiguous and cannot be clarified

on the basis of calculation. This leaves 21 measurements that will be studied further.

The analysis of these measurements is summarized in Table VIII.2. In the table, the first column gives the date of the observation in the Julian calendar and the second column gives the hour, converted to Greenwich mean time. The third column gives the longitude of the moon that we infer from the record; to find this column, we add the fourth and sixth columns in Table VII.2. Finally, the fourth column in Table VIII.2 gives the value of the longitude that we calculate using $\underline{n}_M = -28$ and $\underline{y} = -20$.

The measured and calculated values for -250 March 14 differ by more than 5°. This is far greater than the difference for any other date in the table, so there is probably a scribal error in the measured value. Hence I will omit this observation from the analysis, leaving 20 measurements for final use.

On the average, the observations in Table VIII.2 were made 20.4 centuries before the epoch in 1790 to which the value of $\underline{y}$ is referred. From Equation I.15, this means that the average value of ΔT (dynamical time minus Greenwich mean time) is 13 132 seconds for the observations in the table, if we assume that $\underline{y}$ is exactly -20. Now let λ_m denote the measured longitude of the moon in Table VIII.2 (column 3) and let λ_c denote the calculated value (column 4). If the value $\underline{y} = -20$ were correct, the average value of $\lambda_m - \lambda_c$ would be zero. However, the average value of this quantity from Table VIII.2 is +0°.0266, so the value of -20 needs a small correction.

In particular, we need to increase λ_c by 0°.0266. On the average, the moon moves about 0°.549 per hour, so we need to increase the value of the time used in the calculations by 0.0485 hours. That is, we need to increase the value of dynamical time, according to the observations, by 0.0485 hours or 174 seconds. This means, still according to Equation I.15, that we must change $\underline{y}$ by -0.3. In other words, according to the observations in Table VIII.2, the value of $\underline{y}$ is -20.3. As always, we must attach a standard deviation to this estimate.

When we add 174 seconds to the value of the time in order to increase the value of λ_c, we find that the standard deviation of the residuals in Table VIII.2 becomes 1°.24. Thus the standard deviation to be attached to the value of $\lambda_m - \lambda_c$ given above, namely +0°.0266, is $1°.24/\sqrt{20} = 0°.277$. It takes the moon about 1820 seconds to move this far, so 1820 is the standard deviation to be attached to the value of dynamical time. To change dynamical time by 1820 seconds at the epoch in question takes a change of 2.8 in the value of $\underline{y}$. Hence

$$y = -20.3 \pm 2.8. \qquad\qquad \text{(VIII.5)}$$

I take this to apply at the beginning of the year −250.

In analyzing the lunar conjunctions, we deal only with the longitude of the moon referred, ultimately, to the stars. It is trivial to calculate how the lunar longitude depends upon the acceleration $\underline{\dot{n}}_M$ and thence to calculate how the value of $\underline{y}$ in Equation VIII.5 depends upon $\underline{\dot{n}}_M$. I leave this exercise to the reader in case he should wish to alter the value of $\underline{\dot{n}}_M$.

TABLE VIII.3

ANALYSIS OF THE BABYLONIAN LUNAR ECLIPSES

Date	Quantity[a] observed	Observed value	Calculated[b] value	$\partial \underline{Q}/\partial \underline{y}$	$\partial \underline{Q}/\partial \underline{\dot{n}}_M$
−522 Jul 16	B	$19^h.52$	$19^h.680$	0.234	−0.168
−522 Jul 16	M	0.47	0.568	0	−0.017
−521 Jan 10	B	$−0^h.91$	$0^h.400$	0.235	−0.143
−239 Nov 3	B	$3^h.11$	$3^h.774$	0.180	−0.122
−211 Apr 30	B	$1^h.01$	$0^h.993$	0.176	−0.114

[a]In this column, B denotes the time of the beginning of the eclipse in Greenwich mean time, as inferred from the record. M denotes the magnitude of the eclipse.

[b]Calculated using $\underline{y}$ = −20 and $\underline{\dot{n}}_M$ = −28.

---

4. Babylonian Records of Lunar Eclipses

Table VII.5 lists four Babylonian measurements of the beginning time of a lunar eclipse and one measurement of the magnitude of a partial eclipse. Thus we have five measurements altogether that relate to lunar eclipses, but two of them relate to the same eclipse, so only four individual eclipses are involved.

The analysis of these five measurements is summarized in Table VIII.3. In the table, the first column gives the date of the eclipse and the second column gives the nature of the observation. M in this column means that the magnitude was observed, otherwise the quantity is the beginning time. The third column gives the observed value of the quantity measured. When the observed quantity is a time, this column gives the Greenwich mean time that we infer from the record. I take the time to be negative for −521 January

10 in order to use the date that Oppolzer [1887] assigns.†
The fourth column gives the value of the quantity in question that we calculate using the standard values $\dot{y}$ = -20 and
$\underline{\dot{n}}_M$ = -28. Finally, the last two columns give the partial
derivatives of the quantity in question (Q) with respect to
$\underline{\dot{y}}$ and $\underline{\dot{n}}_M$. Note that $\partial Q/\partial \underline{\dot{y}}$ is identically zero when the
quantity is the magnitude of an eclipse: the magnitude
cannot depend upon the orientation of the earth about its
axis.

From the single observation of the magnitude on -522
July 16, we get $\underline{\dot{n}}_M$ = -22.2, but we cannot get any information about $\underline{\dot{y}}$ from this observation. In order to attach a
standard deviation to this estimate of $\underline{\dot{n}}_M$, in view of the
fact that we have only one measurement, we must look ahead
to Section IX.1, where I study the magnitudes of lunar
eclipses as measured by Islamic astronomers. There I will
find that the standard deviation in the Islamic measurements
is 0.084. This is almost exactly 1 digit. Now, Islamic
astronomers made improvements over the Babylonian astronomers in their measurements of time, but there is no evidence known to me that they made any improvement in the matter of estimating eclipse magnitudes with the naked eye.
Thus I take 0.084 as the standard deviation of the measured
magnitude for -522 July 16. Since the derivative of the
magnitude with respect to $\underline{\dot{n}}_M$ is -0.017 for this eclipse, the
standard deviation in the estimate of $\underline{\dot{n}}_M$ is 0.084/0.017 =
4.9. Hence the measurement of magnitude gives us

$$\underline{\dot{n}}_M = -22.2 \pm 4.9. \hspace{3cm} \text{(VIII.6)}$$

This differs by slightly more than one standard deviation
from the value of -28 that I have adopted in this work.

Now we turn to the measurements of the times when the
eclipses began. If we use the "time bins" defined in Appendix I, we should put the two eclipses in -522/-521 into one
bin and the two eclipses in -239 and -211 into another.
When I divide the eclipses this way, I get almost identical
values for $\underline{\dot{y}}$ from the two pairs, but I have no reliable way
to derive a standard deviation of the values. However,
since the values are almost the same, it should be legitimate to combine all four measurements into a single determination of $\underline{\dot{y}}$.

When I do this, I get $\underline{\dot{y}}$ = -22.7. With this value of $\underline{\dot{y}}$
(and with $\underline{\dot{n}}_M$ = -28), the residuals between the calculated
and measured values of the beginning times range from about
10 minutes to about 40 minutes, and the formal standard
deviation is 0.56 hours, about 34 minutes. In spite of the
small sample, this is a reasonable value. From a large body

---

†Oppolzer assigns his dates on the basis of the time he calculates for the middle of the eclipse.

of similar Islamic measurements in Section IX.2, we will
find a standard deviation of 0.23 hours. It is plausible
that the Babylonian measurements, made more than a millenium
earlier, should have slightly worse than twice this level of
error.

If we accept 0.56 hours as the standard deviation of
the measured times, the standard deviation in the estimate
of $\underline{y}$ is 1.4. Thus the eclipse times in Table VIII.3 give us
the estimate

$$\underline{y} = -22.7 \pm 1.4. \tag{VIII.7}$$

However, the matter is considerably more interesting
than this. We recall that the eclipse of -239 November 3
was recorded as beginning 12 minutes before sunrise. Let us
ignore the time and merely use the condition that the
eclipse was observed in Babylon. It could not have been
observed there if $\underline{y}$ were algebraically as large as -20.[†] As
Table VIII.3 shows, the eclipse began at $3^h.774$, Greenwich
time, if $\underline{y} = -20$, but sunrise that morning was $6^h.32$, local
mean time, which is $3^h.36$, Greenwich mean time. Hence, if $\underline{y}$
= -20, the eclipse began after sunrise (or moonset) in
Babylon and it was not visible there.

From Table VIII.3, we find that the beginning time of
the eclipse is $3^h.774 + 0.180\delta y$. If we equate this to the
time of sunrise, which is $3^h.36$, we get $\delta y = -2.3$ or $\underline{y} =$
-22.3. The value of $\underline{y}$ must be algebraically less than this
if the eclipse was even visible in Babylon. In other words,
the record tells us that $\underline{y} \leq -22.3$ in the year -239.

The eclipse could have begun some time before the measured time, so the record does not give us a tight lower
limit to $\underline{y}$. However, if we combine Equation VIII.7 with the
upper limit just found, we have

$$\underline{y} = -22.7 \begin{array}{c} + 0.4 \\ - 1.4 \end{array}. \tag{VIII.8}$$

The upper limit seems to be a rigid one, with little chance
that it can be exceeded. The lower limit is merely statistical, with about 1 chance in 3 that it can be exceeded. We
may take -373 as the epoch for Equation VIII.8.

If we are indeed going to accept the upper limit in
Equation VIII.8 as a rigid one, we must look carefully at
the credentials of the record dated -239 November 3. So far
as I can see, there are two and only two points we must settle if we are to accept the limit as rigid. These are: (1)
Is the year indeed the year 72 of the Seleucid era? If so,
the eclipse is undoubtedly that of -239 November 3. (2) Was
the eclipse actually observed, or was it merely predicted?

_____

[†]I remind the reader that I am taking $\dot{n}_M = -28$ exactly. If
this value is not correct, some of the numbers in the discussion will need changing, but the main conclusion will
still be correct.

If the year is 72, and if the eclipse was observed, I think we must accept the upper limit in Equation VIII.8 as rigid. Let us look first at the dating of the record.

The record [Schaumberger, 1933] is from a cuneiform text of the sort that Sachs [1948] aptly calls a "goal-year" text. This sort of text applies to a goal-year that we may call Y. A complete goal-year text gives information about the moon and the planets from years that are earlier than the year Y, but the year involved is different for each body. Specifically, the goal-year text gives observations of Saturn for the year Y - 59, of Jupiter for either or both of the years Y - 71 or Y - 83, of Mars for either or both of the years Y - 47 or Y - 79, of Venus for the year Y - 8, of Mercury for the year Y - 46, and of the moon for the year Y - 18.†

The text that concerns us here has lost most of the sections that it presumably contained when it was new, and the surviving part has only three sections. Two of these three sections are badly damaged. The part that is best preserved has observations of Jupiter from the year 19 of the Seleucid Era. A second part has only a single observation of Jupiter from the year 7. We see from the intervals given above that the text must have the goal-year 90 of the Seleucid Era.

The statement of the year has been lost from the third section of the text, which contains references to a solar eclipse and a lunar eclipse. If this text follows the usual practice, the eclipses must be in the year 72 of the Seleucid Era. There were eclipses that year on the dates stated, but the solar eclipse was not visible in Babylon, as Schaumberger points out, and as I have already said in Section VII.7. Since eclipses can occur on only a restricted set of dates in a lunar calendar, this agreement of eclipse dates is not as significant as it would be in a calendar that has lost its relation to the motion of the moon. Nonetheless, I think we are safe in accepting the date of the lunar eclipse as -239 November 3.

It is harder to decide whether or not the eclipse was actually observed. The text first refers to a solar eclipse that was not visible, and we do not know whether the missing part once contained a statement that it was not. Thus, I think, the only indication of actual observation is the statement that the eclipse began on the eastern limb of the moon. This, as I have already said, is something a Babylonian astronomer probably knew by -239, and he could have inserted it into a calculation of an eclipse. However, I do not recall seeing such a remark in any other calculated record, it is superfluous in a calculation, and thus its appearance is a strong indication of observation.

---

†See Newton [1976, pp. 101ff] for a layman's explanation of these particular years. Briefly, they involve periods of near-resonance between the motion of the body in question and the length of the year.

In summary, there is strong evidence that the eclipse of -239 November 3 was observed, but the evidence does not extend to certainty. I will return to the problem of this eclipse in the next section.

5. Babylonian Records of Solar Eclipses

Table VII.5 lists two solar eclipses that were observed in Babylon, those of -321 September 26 and -135 April 15. We have a measured time for both eclipses, and we also have the statement that the eclipse of -135 April 15 was total. I used the statement of totality in volume 1 and I also use it here, in Appendix I, with the circumstances calculated afresh with the DE102 ephemeris. I take up the eclipse of -135 April 15 first.

There are two problems connected with this record. For one thing, we cannot take it for granted that the eclipse was in fact total just because the record says it was. We have found too many instances of an eclipse that could not possibly have been total but that was so recorded, sometimes in a quite picturesque manner.

The second problem concerns the time. As we saw in Section VII.6, we cannot tell from the language whether the eclipse began 96 minutes after sunrise or whether it became total then. I decided to see if we could distinguish between the possibilities on the basis of calculation. Since the time of an eclipse is a sensitive function of $\underline{y}$ for an eclipse this old, I did not expect to succeed. However, we are the beneficiaries of a little serendipity.

Obviously, the time could not be the beginning of totality unless the eclipse was actually total in Babylon. It turns out that the eclipse could not have been total unless $\underline{y}$ lies in the narrow range from -21.2 to -23.5. But, if $\underline{y}$ does lie in this range, the recorded time must be the beginning of the eclipse and not the beginning of totality.†

That is, the recorded time is the beginning of the eclipse if the eclipse were total. If the eclipse were not total, the recorded time is a fortiori the time of the beginning. Thus the recorded time is the time of the beginning whether the eclipse was total or not.‡

I estimate that sunrise on -135 April 15 in Babylon was at 5.58 hours, local mean time. Since the eclipse began 96 minutes later, the measured beginning was at 7.18 hours,

---

†We remember from Section VII.6 that one record specifically says that "there was complete night" at 96 minutes after sunrise. This record cannot refer to -135 April 15 unless there is an error in the translation.

‡To sum up the situation, we know that the recorded time applies to the beginning of the eclipse, but we do not know whether the eclipse was total or partial.

local time, or 4.22 hours, Greenwich time. From calculating
the circumstances of the eclipse as a function of y and $\underline{\dot{n}}_M$,
I find that the beginning time $\underline{B}$ is

$$\underline{B} = 5^h.215 + 0^h.171y - 0^h.104\underline{\dot{n}}_M.$$

If we set $\underline{\dot{n}}_M = -28$, set $\underline{B} = 4.22$ hours, and solve for y, we
get -22.8. If we use 0.56 hours as the standard deviation
of the measured time, the standard deviation of y is 3.3.
Thus the measured time of the eclipse of -135 April 15 gives
us

$$y = -22.8 \pm 3.3. \qquad\qquad (VIII.9)$$

We will find out in Section X.1 that y was close to
-22.4 in the year -135, so the eclipse was actually total.
Further, Mercury, Venus, Mars and Jupiter were all above the
horizon while Saturn was not. Thus it turns out that the
record is accurate in its details, but we know this only a
posteriori. We may not conclude from the record itself that
the eclipse was total, not even with the details about the
planets. These are details that the Babylonian astronomers
would have known without observation.

Now let us turn to the eclipse of -321 September 26.
According to Table VII.5, this record says that the eclipse
began 12 minutes before sunset. I calculate that sunset in
Babylon on this date came at 15.02 hours, Greenwich mean
time, so the recorded time is equivalent to 14.82 hours,
Greenwich time. The calculated time $\underline{B}$ of the beginning is
given by

$$\underline{B} = 13^h.178 + 0^h.215y - 0^h.228\underline{\dot{n}}_M. \qquad (VIII.10)$$

If we equate B in Equation VIII.10 to the measured time,
taking $\underline{\dot{n}}_M = -28$, we get $y = -22.1$. If we equate it to the
time of sunset, we get a value that is 0.9 larger; and if we
equate it to half an hour before the measured time, we get a
value that is 2.3 smaller. Thus the record gives us

$$y = -22.1 \begin{array}{l} + 0.9 \\ - 2.3 \end{array}. \qquad\qquad (VIII.11)$$

If we can be certain that the year of the eclipse is
-321, and if we can be certain that the eclipse was actually
observed and not calculated, we can take the upper limit in
Equation VIII.11 as a rigid one. The date of the eclipse
(Section VII.6) is given in full in the record, and it
agrees exactly with the date -321 September 26. Hence I
think the date is established beyond a reasonable doubt.

I have not seen the full text, only the parts that were
quoted in the references mentioned in Section VII.6. However, the record is described as part of an astronomical
diary, and it records the weather on the day of the eclipse.
Since a diary usually records observations, not calculations, and since it refers to the weather, we have strong
but not certain evidence that the eclipse was actually
observed.

If we assume that both this eclipse and the lunar eclipse of -239 November 3 were actually observed, we must give preference to the latter record (Equation VIII.8), since the upper limit in it is smaller than the upper limit in Equation VIII.11. Since we cannot be entirely certain that either eclipse was observed, I think the conservative course is to use both Equations VIII.8 and VIII.11, and to treat the upper limits given in both equations as reflecting standard deviations rather than rigid limits. Hence we will have separate standard deviations for positive and for negative deviations for the data from Babylonian astronomy.

TABLE VIII.4

THE ECLIPSE OF -88 SEPTEMBER 29[a]

| $\dot{n}_M$ | y | Calculated hour, local apparent time | | | Magnitude |
		Middle	End	Midpoint	
-28	-20	16.326	17.438	16.882	0.895
-28	-25	15.360	16.581	15.970	0.952
-23	-20	15.651	16.835	16.243	0.877

[a]Calculated for the point at latitude 34°.06, longitude 113°.51.

6.   The Eclipse of -88 September 29

We presented a Chinese record of this eclipse in Section VI.4. There we decided that the observation was made somewhere within a rectangle whose center is at latitude 34°.06 and longitude 113°.51. The rectangle extends 1°.15 in latitude in each direction from the center, and 3°.57 in longitude in each direction. We decided that the uncertainty in position introduces an uncertainty in the measured times that has a standard deviation of 0.22 hours, and that the statement of the time (including the likely error in measuring the time) introduces a further uncertainty of 0.45 hours. The resultant standard deviation is 0.50 hours.

The record says that the eclipse was not quite total, that it ended before sunset, and that the midpoint between the middle of the eclipse and its end was at what we call 16 hours, local apparent time.

Table VIII.4 summarizes the calculated circumstances of the eclipse, for each of the three combinations of y and of $\dot{n}_M$ that were used in the computations, for the center of the rectangle. The first column gives the value of $\dot{n}_M$ that was used and the second column gives the value of y. The third column gives the time of the middle of the eclipse, in local apparent time at the center of the rectangle, and the fourth column gives the time of the end. The fifth column gives the local apparent time of the midpoint between the middle of the eclipse and the end, still for the center of the rectangle.

As a matter of interest, the sixth column gives the magnitude that is calculated for the center of the rectangle. We know that the eclipse was not quite total, but it would have been if y had been much less (algebraically) than -25. Thus we can get a moderately useful lower limit to y (for a given value of $\underline{\dot{n}}_M$) from the statement that the eclipse was not total.

It is not clear from the table, but the record also furnishes an upper limit to y. The record says that the eclipse ended before sunset, but the sun would have set still partially eclipsed if y had been much larger (algebraically) than -20. Thus this single record gives us both upper and lower limits to the value of y that are reasonably close together. In other words, we can put useful limits to y just from the qualitative conditions in the record.

However, we can do somewhat better if we use the quantitative condition, namely that the point midway between the middle and the end of the eclipse was at 16 hours, local apparent time. From the numbers given in Table VIII.4, the reader should have no trouble in deriving the equation of condition:

$$17.153 + 0.1824\underline{y} - 0.1208\underline{\dot{n}}_M = 16 \pm 0.5.$$

If we take $\underline{\dot{n}}_M = -28$ as usual, this gives

$$y = -24.8 \pm 2.7. \qquad\qquad\qquad \text{(VIII.12)}$$

7.   <u>Observations from Greek Astronomy</u>

In Chapter VI, we found only three usable records of lunar observations from all of Greek astronomy. All three are from a time far after the flowering of Greek scholarship.

Proceeding chronologically, we come first to the solar eclipse of 364 June 16, which was observed in Alexandria. According to Table VI.1, the eclipse began at 2 5/6 hours, the middle was at 3 4/5 hours, and the end was at 4 1/2 hours, all in equal hours after apparent noon. The equation of time, in the sense of apparent minus mean, was +0.0538 hours. Hence the beginning was at $14^h.780$, local mean time, and so on.

Table VIII.5 summarizes the analysis of the record. The first column gives the phase of the eclipse in question and the second column gives the measured hour of that phase, in local mean time. The third column gives the hour that we calculate using $\underline{y} = -20$ and $\underline{\dot{n}}_M = -28$, and the next two columns give the partal derivatives of the hour with respect to $\underline{y}$ and to $\underline{\dot{n}}_M$, respectively. The last column gives the

TABLE VIII.5

THE ECLIPSE OF 364 JUNE 16

Phase	Measured local mean time $\underline{h}$	Calculated local mean time[a] $\underline{h}$	$\partial \underline{T}/\partial \underline{y}$	$\partial \underline{T}/\partial \underline{\dot{n}}_M$	Weight
Beginning	14.780	15.590	+0.0985	−0.0844	4
Middle	15.746	16.502	+0.0987	−0.0783	1
End	16.446	17.316	+0.0998	−0.0736	4

[a]Using $\underline{y}$ = −20 and $\underline{\dot{n}}_M$ = −28.

weight to be attached to each measurement; as I said in Section VI.1, the middle of an eclipse cannot be measured with the same precision as the beginning and end, so I arbitrarily decided to give the middle time one fourth of the weight of the other two times.

It is interesting to look at the middle phase. The measured time of the middle is greater than the average of the beginning and end by 8 minutes, as I have already pointed out, which shows that the middle was separately observed. Further, the calculated middle is later than the average of the calculated beginning and end, but only by about 3 minutes. This increases the likelihood that the middle was observed, but I know of no clue to the method of observation.

When I find a value of y by the method of weighted least squares, I get −28.4. We cannot base the standard deviation of y upon the residuals because the errors are surely not independent. Instead, I take the standard deviation in the measured times to be half an hour, which is about the value I used with some of the Babylonian eclipses. This leads to 5.0 as the standard deviation of y. Thus the eclipse of 364 June 16 gives

$$y = -28.4 \pm 5.0. \tag{VIII.13}$$

Now we may turn to the conjunctions of the moon observed by Heliodorus and his brother Ammonius in the early sixth century. These observations are summarized in Table VI.1.

Before we infer a value of y from these observations, let us first look at the latitude (or declination) of the moon that is implied by the observations. On 503 February 21, the record says that Saturn appeared to bisect the moon.

-190-

However, in both latitude and declination the center of the moon was about 0°.15 below Saturn. In other words, instead of bisecting the moon, Saturn barely passed below its upper limb. This does not speak well for the quality of the observation.

TABLE VIII.6

TWO LATE GREEK MEASUREMENTS OF THE LUNAR LONGITUDE

Date	Hour[a]	Measured longitude °	Calculated longitude[b] °	$\partial\lambda/\partial\underline{y}$	$\partial\lambda/\partial\underline{\dot{n}}_M$
503 Feb 21	21.205	95.648	95.544	−0.0400	+0.0302
509 Mar 11	16.881	49.764	49.876	−0.0369	+0.0300

[a]In Greenwich mean time.
[b]Calculated with $\underline{y}$ = −20 and $\underline{\dot{n}}_M$ = −28.

On 509 March 11, the record says that α Tauri was half a degree west of the lunar limb. Since the moon was only a few days old at the time, α Tauri was west of the bright limb, so I take it to be 0°.513 west of the true limb. However, in both declination and latitude, the star was about half a degree south of the center of the moon. Thus the star was really southwest of the moon, rather than west as the record suggests.

On 509 March 11, the lunar distance was somewhat greater than average, so I take its apparent radius to be 15'. Thus I take its longitude to be 0°.763 greater than that of α Tauri. The time given for 503 February 21 is the observers' estimate of the time when the moon and Saturn had the same apparent longitude.

The analysis of the records is summarized in Table VIII.6. The first column gives the date of an observation and the second column gives the hour, reduced to Greenwich mean time. The third column gives the longitude that we infer from the measurement and the fourth one gives the longitude that we calculate using the stated parameters. The last two columns give the customary partial derivatives.

The value of $\underline{y}$ that we infer is −20.0. Since there are only two observations, I do not think we should calculate the standard deviation of the measurements from them. Judging from the accuracy of the latitudes, as well as from the general level of accuracy of Greek observations, we should take something like 15' for the accuracy. This leads to 4.6 for the standard deviation of $\underline{y}$. That is,

$$\underline{y} = -20.0 \pm 4.6 \qquad\qquad \text{(VIII.14)}$$

is the estimate of $\underline{y}$ we make from Table VIII.6. The epoch of this estimate is the year 506.

CHAPTER IX

ANALYSIS OF THE ISLAMIC OBSERVATIONS

1.   The Magnitudes of Lunar Eclipses

The only Islamic observations we have that involve the moon[†] are observations of eclipses, both lunar and solar. For both kinds of eclipse, we have numerous observations of both the magnitudes and times, so we have four classes of observation altogether to analyze. I will start with mea- surements of the magnitudes of lunar eclipses.

Table V.2 summarizes the Islamic measurements of lunar eclipses that are available for this study. We have records of 21 eclipses, with two independent records, called 901 August 3a and 901 August 3b, of the eclipse of 901 August 3, for a total of 22 records. The maximum magnitude was mea- sured in 12 of these records. For a 13th record, namely that of 979 May 14, we cannot tell from the reading whether the recorded magnitude is the maximum or the value at sun- rise. Calculation shows that the moon rose eclipsed in Cairo and that the magnitude seen there decreased steadily from moonrise on. Hence the recorded magnitude, though it was the maximum magnitude visible in Cairo, was not the true magnitude of the eclipse.

TABLE IX.1

ISLAMIC MEASUREMENTS OF THE MAGNITUDES OF LUNAR ECLIPSES

Date	Observed magnitude	Calculated magnitude[a]	$10^4 \times d\underline{m}/d\underline{\dot{n}}_M$
854 Feb 16	0.83	0.915	+33
856 Jun 22	0.71	0.591	−32
883 Jul 23	0.88	0.941	+30
901 Aug 3a	0.96	1.057	+29
901 Aug 3b	0.96	1.057	+29
923 Jun 1	0.78	0.657	+28
927 Sep 14	0.29	0.211	+26
979 Nov 6	0.80	0.839	−25
981 Apr 22	0.25	0.174	+27
981 Oct 16	0.42	0.357	+24
986 Dec 19	0.83	0.916	−27
990 Apr 12	0.63	0.741	−27

[a]Calculated with $\underline{\dot{n}}_M = -28$.

[†]Exclusive of values from tables.

For a reason that will appear later, the analysis of the record of 979 May 14 cannot be readily combined with that of the other measurements of magnitude, and it will be taken up separately. Hence I start with the measurements of the true maximum magnitude. The twelve records that give a measurement of the true maximum are summarized in Table IX.1. The first column in the table gives the date of the eclipse, with two records shown for 901 August 3. Since we are speaking of the maximum magnitude, the hour is not relevant and is not given. The second column gives the recorded magnitude.

The third column gives the calculated magnitude. We recall from Section III.4 that the magnitude of a lunar eclipse does not depend upon the value of y. The values in the third column were calculated with $\dot{n}_M = -28$, while the fourth column gives the derivative of the magnitude m with respect to $\dot{n}_M$.

The value of $\dot{n}_M$ that we infer from Table IX.1 is -26.6. When we recalculate the magnitudes using this value, we find that the standard deviation of the residuals is about 0.08, in good agreement with other results, and the corresponding standard deviation of $\dot{n}_M$ is 8.6.

Now let us look at the two records of the eclipse of 901 August 3, which have concerned me ever since I first studied them [Newton, 1970, pp. 220 and 235]. One of these records is from Antakya and the other is from Raqqa, and both were used by al-Battani [ca. 925]. Both records say that the moon was eclipsed by a little less than its diameter, although the standard deviation in judging the magnitude of an eclipse is about a digit. However, the moon was not eclipsed by a little less than its diameter, which I took to mean 11 1/2 digits. It was totally eclipsed, and by a fairly safe margin. I am suspicious of the validity of these records, and I will find additional reason for suspicion in the next section.

If I omit these records, I change the inferred value of $\dot{n}_M$ by less than a standard deviation. Further, the value of $\dot{n}_M$ derived from Table IX.1 will not have much effect on our final conclusions because of its large standard deviation. Thus I include these records in the analysis in spite of my suspicions and get

$$\dot{n}_M = -26.6 \pm 8.6. \qquad\qquad (IX.1)$$

The epoch of this value is the average of the dates in Table IX.1. This average is 926.

Now we turn to the record of 979 May 14, which records that the magnitude at moonrise was 0.71. The magnitude of a lunar eclipse at a specific time (rather than at the maximum) depends upon both y and $\dot{n}_M$. Further, the time itself in this case also depends upon both accelerations. Thus

TABLE IX.2

ISLAMIC MEASUREMENTS OF THE TIMES OF LUNAR ECLIPSES

Date	Phase[a]	Measured hour[b]	Calculated hour[c]	$\partial T/\partial y$	$-\partial T/\partial \dot{n}_M$
854 Feb 16	B	19.344	19.529	0.0384	0.0364
854 Aug 12	$\overline{B}$	0.009	0.011	0.0383	0.0313
856 Jun 22	$\overline{B}$	0.422	0.377	0.0381	0.0290
883 Jul 23	$\overline{M}$	17.635	17.164	0.0359	0.0298
901 Aug 3a	$\overline{M}$	0.987	0.526	0.0349	0.0289
901 Aug 3b	M	1.044	0.526	0.0349	0.0289
923 Jun 1	$\overline{M}$	17.702	17.521	0.0329	0.0269
	$\overline{E}$	18.936	18.895	0.0329	0.0247
925 Apr 11	$\overline{B}$	16.320	16.531	0.0329	0.0262
	$\overline{E}$	19.790	19.845	0.0329	0.0287
927 Sep 14	B	0.903	1.134	0.0327	0.0299
929 Jan 27	$\overline{B}$	21.318	21.171	0.0325	0.0262
933 Nov 5	$\overline{B}$	1.351	1.414	0.0320	0.0328
979 May 14	$\overline{E}$	17.827	17.796	0.0289	0.0271
979 Nov 6	B	20.073	20.074	0.0295	0.0228
	$\overline{E}$	23.293	23.151	0.0289	0.0268
980 May 3	$\overline{B}$	-1.635	-1.269	0.0287	0.0285
	$\overline{E}$	2.546	2.554	0.0287	0.0278
981 Apr 22	B	1.391	1.464	0.0287	0.0358
	$\overline{E}$	3.061	3.143	0.0288	0.0238
981 Oct 16	$\overline{T}$	2.119	2.019	0.0286	0.0259
983 Mar 1	$\overline{E}$	25.545	25.355	0.0285	0.0252
986 Dec 19	T	2.485	2.565	0.0282	0.0279
	$\overline{A}$	2.851	2.578	0.0281	0.0280
990 Apr 12	$\overline{B}$	19.679	20.181	0.0280	0.0270
	$\overline{E}$	23.150	23.351	0.0280	0.0314
1001 Sep 5	E	17.975	17.690	0.0271	0.0301
1002 Mar 1	$\overline{B}$	21.643	21.656	0.0273	0.0222
1019 Sep 16	$\overline{A}$	21.644	21.673	0.0260	0.0253
$\underline{m} = 0.04$		24.988	24.806	0.0260	0.0293

[a] In this column, B denotes the beginning, M denotes the middle, and E denotes the end of an eclipse. T denotes the phase that Caussin translates as attouchement, which I equate to the beginning. A denotes the phase I call "appearance". Quantitatively, I take this to be the time when the magnitude reaches 0.01.

[b] Converted to Greenwich mean time.

[c] Calculated with $\underline{y} = -20$ and $\dot{\underline{n}}_M = -28$.

the record of 979 May 14, unlike the records in Table IX.1, gives us an equation of condition that involves both y and $\dot{n}_M$:

$$0.71 = 0.950 + 0.0290y - 0.0109\dot{n}_M. \qquad (IX.2)$$

If we take $\dot{n}_M = -28$ and solve for y, we find

$$y = -18.0 \pm 2.4. \qquad (IX.3)$$

The standard deviation in Equation IX.3 comes from taking the standard deviation of the measured magnitude to be 0.07.

2.   The Times of Lunar Eclipses

As I remarked before, Table V.2 presents 22 records of lunar eclipses, with two of them referring to the same eclipse of 901 August 3. Each record contains the measured time for some phase of the eclipse, and some records contain more than one measurement of time. Altogether, we have thirty useable measurements of time associated with lunar eclipses. The analysis of these times is summarized in Table IX.2.

In the table, the first column gives the date of the eclipse and the second column identifies the phase whose time was measured. In this column, B identifies the beginning, M identifies the middle, and E identifies the end, of an eclipse. T identifies the phase that Caussin, in his translation of ibn Yunis [1008], calls attouchement, or touching. For reasons I discussed in Section V.3, I take this to be the same as the beginning. A identifies the phase I called "appearance" in Section V.3, which I take to be the condition when the magnitude equals 0.01.

The third column of Table IX.2 gives the measured time of an observation, reduced to Greenwich mean time, and the fourth column gives the time that we calculate for the phase in question, assuming as usual that $y = -20$ and $\dot{n}_M = -28$. The last two columns give the derivatives of the time with respect to y and to $\dot{n}_M$. Since the derivative with respect to $\dot{n}_M$ is always negative, I tabulate the negative of the derivative for convenience.

Table V.2 contains many more entries than Table IX.2. There are several reasons for this shrinkage. As I explained in Section V.5, a number of the elevations of stars used to measure the time are either ambiguous or impossible, so we cannot determine what the measured time was; examples are the phase A for 983 March 1 and the phase E for 1002 March 1, just 19 years later.

For another reason, the time of an observation was often given in several different ways in Table V.2. In these cases, I have given a single time in Table IX.2. Sometimes, in Table V.2, the observer gave an observation from which the time was determined as well as a value of the time that he calculated from the observation. In these cases, I have given only the time that I calculate from modern theory. In other cases, the observer gave several different measurements from which he determined the time, such as the positions of several stars. In these cases, I have calculated the time corresponding to each observation and have given the average in Table IX.2.

The eclipse of 923 June 1 poses a problem that could probably be resolved with enough effort. Part of the record says that the moon rose eclipsed, and that the magnitude was 0.27 at moonrise. However, for some of the combinations of $\underline{y}$ and $\underline{\dot{n}}_M$ that I used in the calculations, including those used to find the derivatives, the moon rose eclipsed and for others it rose before the eclipse started. It does not seem worth the effort to decide what actually happened, and I have simply omitted this phase of the eclipse from Table IX.2.

The value of $\underline{y}$ that we infer from Table IX.2 is -18.8, the standard deviation of the residuals is 0.23 hours, and the corresponding standard deviation of $\underline{y}$ is 1.3. Thus:

$$\underline{y} = -18.8 \pm 1.3. \tag{IX.4}$$

When I assigned the records of solar eclipses to "time bins" in volume 1 (and in Appendix 1 in this volume), I made a division between bins at 900. Four of the observations in Table IX.2 come before this date and the remaining 26 observations come after it. Since there are so few observations before 900, it is not worthwhile to separate them from the others, so I will consider all the observations in Table IX.2 as belonging to a single set, with the average epoch of 952.

During 0.23 hours, which is the standard deviation of the residuals, the magnitude typically changes by about 0.2. This is far greater than the standard deviation in estimating the magnitude, as we saw in the preceding section. Further, when we plot the distribution of the residuals, we see that they look normal for errors up to about 0.25 hours, that there is then a gap up to 0.4 hours and that there are five errors greater than 0.4 hours. Two of these errors are for the beginnings of the eclipses of 980 May 3 and 990 April 12. We saw in Section V.5 that there were scribal errors in recording the observations from which the times were deduced for both occasions, and I adopted an emended reading suggested by Knobel [1879] on the basis of his knowledge of Islamic numerals. I believe we are safe in concluding that Knobel's readings are not correct.

Two other large errors are for the two records of the eclipse of 901 August 3 that were used by al-Battani [ca. 925]; the observation labelled 901 August 3b was in fact made by him. We remember from the preceding section that both records say that the eclipse was partial and that they give identical values of the magnitude, although the eclipse was in fact total. They also give large but nearly equal errors in the recorded times. I think we are safe in saying

that there has been some distortion in these records and
that we should omit them from the analysis.

The remaining large error is for the eclipse of 883
July 23. This observation was made by al-Battani, and he
simply states the time of the middle of the eclipse, without
telling us how he measured it. I will omit this observation
also. It is interesting that three of the five largest errors in the eclipse times come from the work of al-Battani,
and two of them were made by him.

When we omit the five records just discussed and make a
new analysis, we find

$$\underline{y} = -19.7 \pm 0.9. \hspace{3cm} (IX.5)$$

When we use Equation IX.5 instead of Equation IX.4, the
epoch changes slightly, to 948. The standard deviation of
the residuals falls from 0.23 hours to 0.15 hours. During
this time, the magnitude changes by about 0.12 in a typical
case. This is still greater than the standard deviation in
judging the magnitude, so it implies that the errors in the
times are dominated by the errors in measuring the time, not
by errors in judging the phase.

The standard deviation of 0.15 hours is the same as 9
minutes. I used this value in Section IV.2 above, on the
basis of the results obtained here.

3.   The Magnitudes of Solar Eclipses

Table V.1 summarizes the Islamic records of solar
eclipses that are available for this work. There are records of 13 eclipses, with two independent records of the
eclipse of 901 January 23, for a total of 14 records. None
of the eclipses was total, and the magnitude is recorded in
12 of the records. In one of these records, namely that of
928 August 18, it is not clear whether the magnitude recorded is the maximum magnitude or the magnitude at sunrise.
The calculations indicate that the maximum magnitude is
probably meant, although the recorded value is considerably
too small with either interpretation. I will take the recorded value to be that of the maximum for the sake of definiteness; the final results are almost the same with either
interpretation.

Similarly, for the eclipse of 979 May 28, it is not
clear from the record whether the recorded magnitude is the
maximum or the value at sunset. For this record, it turns
out that the recorded magnitude is both: the sun set
eclipsed, and the magnitude was still increasing at sunset.

The analysis of the magnitudes is summarized in Table
IX.3. In the table, the first column as usual gives the
date of the eclipse. The second column gives the observed
magnitude, while the third column gives the magnitude calcu-
lated using $\underline{y} = -20$ and $\underline{n}_M = -28$. The last two columns
give the partial derivatives of the magnitude $\underline{m}$ with respect
to $\underline{y}$ and to $\underline{n}_M$, respectively.

The first record in the table, that of 792 November 19, is not an Islamic record at all. It is the single measurement of the magnitude of a solar eclipse that we have from China which seems to have been made by a trained astronomer. I have included it with the Islamic measurements for convenience.

TABLE IX.3

ISLAMIC MEASUREMENTS OF THE MAGNITUDES OF SOLAR ECLIPSES

Date	Measured magnitude	Calculated magnitude[a]	$\partial m/\partial y$	$\partial m/\partial \dot{n}_M$
792 Nov 19[b]	0.31	0.235	+0.0102	−0.0109
866 Jun 16	0.62	0.661	−0.0025	−0.0022
891 Aug 8	0.76	0.857	−0.0088	+0.0116
901 Jan 23a	0.63	0.679	+0.0028	−0.0055
901 Jan 23b	0.70	0.690	+0.0026	−0.0053
923 Nov 11	0.75	0.785	+0.0063	−0.0074
928 Aug 18	0.34	0.223	−0.0098	+0.0047
977 Dec 13	0.65	0.608	−0.0058	+0.0071
978 Jun 8	0.46	0.511	−0.0037	+0.0008
979 May 28	0.46	0.407	−0.0103	+0.0013
985 Jul 20	0.25	0.288	−0.0042	+0.0065
993 Aug 20	0.73	0.951	+0.0013	+0.0019
1004 Jan 24	0.92	0.988	−0.0046	−0.0050

[a]Calculated with $y = -20$ and $\dot{n}_M = -28$.

[b]This is an isolated Chinese record that is included here for convenience.

When we compare the measured and calculated magnitudes in Table IX.3, we see that all the measured values look reasonable except for 993 August 20. The measured magnitude for this date is far too low, so there is probably a scribal error in the record. I omit this record from the analysis.

When we proceed in the usual way to find y from the other records, the result is

$$y = -19.8 \pm 2.8. \tag{IX.6}$$

The standard deviation of the residuals is 0.066. The epoch for Equation IX.6 is 932.

I will comment on the records 901 January 23a and 901 January 23b, which are from the work of al-Battani [ca. 925], in Section IX.5.

TABLE IX.4

ISLAMIC MEASUREMENTS OF THE TIMES OF SOLAR ECLIPSES

Date	Phase[a]	Measured hour[b]	Calculated hour[c]	$\partial T/\partial y$	$-\partial T/\partial \dot{n}_M$
829 Nov 30	B	4.601	4.027	0.0477	0.0419
	$\overline{\text{E}}$	6.447	6.298	0.0482	0.0390
866 Jun 16	$\overline{\text{B}}$	9.570	9.399	0.0595	0.0447
	$\overline{\text{M}}$	10.716	10.748	0.0547	0.0447
	$\overline{\text{E}}$	11.981	12.015	0.0481	0.0419
891 Aug 8	M	10.573	10.349	0.0503	0.0497
901 Jan 23a	$\overline{\text{M}}$	6.178	5.941	0.0441	0.0344
901 Jan 23b	$\overline{\text{M}}$	6.151	5.986	0.0452	0.0353
923 Nov 11	$\overline{\text{M}}$	4.369	4.300	0.0363	0.0269
	$\overline{\text{E}}$	5.538	5.335	0.0412	0.0315
928 Aug 18	E	3.499	3.399	0.0374	0.0301
977 Dec 13	$\overline{\text{A}}$	6.301	6.272	0.0395	0.0301
	$\overline{\text{E}}$	8.620	8.620	0.0417	0.0302
978 Jun 8	$\overline{\text{A}}$	12.376	12.123	0.0472	0.0480
	$\overline{\text{E}}$	14.707	14.755	0.0294	0.0345
979 May 28	A	16.221	16.130	0.0249	0.0266
985 Jul 20	$\overline{\text{B}}$	14.941	14.619	0.0358	0.0346
	$\overline{\text{E}}$	16.305	16.046	0.0229	0.0184
993 Aug 20	$\overline{\text{B}}$	5.598	5.658	0.0348	0.0272
	$\overline{\text{M}}$	7.006	6.848	0.0391	0.0305
	$\overline{\text{E}}$	8.280	8.170	0.0430	0.0336
1004 Jan 24	A	14.035	13.754	0.0317	0.0272
	$\overline{m} = \tfrac{1}{4}$	14.172	14.035	0.0309	0.0267
	$\overline{m} = \tfrac{1}{2}$	14.617	14.319	0.0302	0.0259
	M	15.051	14.863	0.0284	0.0247
1019 Apr 8	$\overline{m} = 0.39$	1.618	1.853	0.0263	0.0247

[a] The symbols in this column have the same meaning as in Table IX.2.

[b] Converted to Greenwich mean time.

[c] Calculated with $y = -20$ and $\dot{n}_M = -28$.

---

4. The Times of Solar Eclipses

For each of the 14 records of solar eclipses in Table V.1, we have a measurement of the time of at least one phase of the eclipse, and in one case (1004 January 24) we have the times of four phases. Altogether, we have 26 measurements of time in the 14 records. These measurements are summarized in Table IX.4.

In the table, the first column gives the date of the eclipse and the second column identifies the phase whose

time was measured. The letters in this column have the same
meanings as they did in Table IX.2. (However, the phase
marked $\underline{T}$ in Table IX.2 does not happen to occur in Table
IX.4). The third column gives the measured time of the
phase specified in the second column, but converted to
Greenwich mean time, while the fourth column gives the calculated hour of the phase in question, also in Greenwich
mean time. This time is calculated using $\underline{y}$ = -20 and $\underline{\dot{n}}_M$ =
-28. Finally, the last two columns give the partial deriva-
tives of the calculated time $\underline{T}$ with respect to $\underline{y}$ and to $\underline{\dot{n}}_M$.

When we solve for $\underline{y}$ from the observations in Table
IX.4, we find -16.5 ± 0.8, and the standard deviation of the
residuals is $0^h.158$. Thus, we have a standard deviation of
about 9 minutes for the measured times of both lunar and
solar eclipses. When we look at the individual residuals,
they all appear consistent except the first and last, which
do not seem to belong to the same population as the others.
When we omit these two, we find $\underline{y}$ = -16.8 ± 0.6, with the
standard deviation of the residuals being $0^h.124$. The differences do not seem statistically significant, so I choose
the solution that uses all the data, namely

$$\underline{y} = -16.5 \pm 0.8, \tag{IX.7}$$

even though I believe that the first and last observations
contain scribal errors. The epoch of this solution is 941.

5. The Observations Used by al-Battani

al-Battani [ca. 925] used three records of solar eclipses and three of lunar eclipses, and he made two of each
type himself in Raqqa. The ones he made himself are 891
August 8 and 901 January 23b of the solar eclipses, and 883
July 23 and 901 August 3b of the lunar eclipses. The ones
he used but did not make himself are 901 January 23a of the
solar eclipses and 901 August 3a of the lunar eclipses.

I have already mentioned that I find the records of the
lunar eclipses to be suspicious. To start with, both
records of the lunar eclipse of 901 August 3 say that the
eclipse was slightly less than total, although the eclipse
was definitely total and the period of totality lasted
[Oppolzer, 1887] 34 minutes. Further, the errors in the
recorded times, though they agree well with each other, are
so large that they must be errors in recording and not in
observation.

Still further, the error in the time that al-Battani
measured for the eclipse of 883 July 23 is so large that it
must also have been a recording error. In all, al-Battani
measured four quantities associated with lunar eclipses, and
all four quantities, though agreeing well with each other,
have such large errors that they must be recording errors
and not genuine errors of observation. In addition, of the
two quantities that al-Battani reports but did not measure
himself, the time in the record 901 August 3a must also have
a recording error that happens to agree well with the other

recording errors. Only the magnitude in the record 901
August 3a, of all the six quantities in question, has a reasonable error.

It is quite unlikely that there should be recording
errors in five out of six measurements, much less errors
that agree well with each other. If we had only the lunar
eclipses to consider, we would be entitled to a suspicion
that the records had deliberately been forced to agree with
some preconceived theory. We have a parallel in Islamic
astronomy in the work of Abu Sahl al-Kuhi (see Newton
[1972a] for a discussion). al-Kuhi claims that, around the
year 988, he carefully measured the obliquity of the ecliptic and found it to be exactly 23° 51' 20". This is exactly
the value that Ptolemy claimed to have measured in about the
year 140, and that is attributed to Eratosthenes in about
-250.

I think there is little doubt that al-Kuhi fabricated
his measurement of the obliquity, and I was anticipated in
this conclusion by al-Biruni [1025] almost a millenium ago.

al-Battani also reports both the magnitude and the time
of maximum eclipse for the solar eclipse records of 891 August 8, 901 January 23a, and 901 January 23b, and he made the
measurements of the first and last of these records himself.
The errors in these measurements show no trace of suspicious
circumstances: the errors are all of a normal size and they
show no particular correlation with each other. The evidence of the solar eclipses contradicts the evidence of the
lunar ones.

I can reach no conclusion in the matter of al-Battani's
records. His records of solar eclipses look perfectly
straightforward while his records of lunar eclipses show
strong evidence of being forced to agree with a preconceived
theory. I can only leave this as a matter for further
study.

6.   Suggested Corrections to Some Data from ibn Yunis

Almost every ancient or medieval source of astronomical
data contains some large errors of the type I have called
recording errors. In some cases the errors are accidental
ones made by the original recorder of the data, and in some
cases the errors were made by later copyists.† Knobel
[1879] studied some recording errors found in the work of
ibn Yunis [1008] and suggested what the original recordings
were, on the assumption that the errors were accidentally
introduced by copyists.

Specifically, Knobel assumed that a copyist confused
one symbol with another that resembles it and that he put

---

†We also know of two copyists who deliberately introduced
  errors into the data. See Newton [1977], and Sections
  III.2 and VI.2 above.

down the wrong symbol as a result. To give an analogy in modern writing, the script forms of three and eight are quite a bit alike. If the left side of an "8" happens to be faint, a copyist can misread it and copy it as a "3". Con-

TABLE IX.5

SOME TEXTUAL VARIANTS IN CAUSSIN'S EDITION OF IBN YUNIS

Date	Caussin's reading		Knobel's reading		Correct reading
	Object	Altitude	Object	Altitude	
929 Jan 27	α Boo	18°	α Boo	33°	Knobel
980 May 3	Moon	47° 40'	Moon	41° 40'	Neither
983 Mar 1	Moon	66°	Moon	62°	Neither
986 Dec 19	Moon	50° 30'	Moon	28° 30'	Knobel
990 Apr 12	Moon	38°	Moon	33°[a]	Neither
1002 Mar 1	α Boo	52°	α Vir	12°	Caussin
	α Aur	14°	α Tau	14°	Caussin
	α Boo	35°	α Vir	35°	Neither
1004 Jan 24	Sun	18° 30'	Sun	17° 30'	b

[a]This emendation was made not by Knobel but by an anonymous scholiast.

[b]The indicated times are so close together that we cannot tell which reading is correct.

versely, if the open ends of a "3" happen to close more than usual, it can easily be misread as an "8".† Similarly, in some early Greek writing, in which the letters of the alphabet did double duty as numerals, an alpha, which represented "1", was written as a triangle with one side extended, instead of being written like our modern A. This could easily be confused with Δ, which represented "4".

Table IX.5 summarizes the emendations suggested by Knobel, as well as one (for 990 April 12) that was suggested by an unknown scholar in a marginal note in the copy of ibn Yunis that I used. The first column in the table gives the date of the record involved. The second column gives Caussin's reading of the body used in determining the time (Caussin is the editor of the published version of ibn Yunis), and the third column gives his reading of the altitude of that body. The fourth and fifth columns give the changes that Knobel suggests and that were discussed in Sections V.3, V.4, and V.5. The sixth column tells which reading, if either, proves to be correct on the basis of calculation.

†This confusion is not likely with some printed forms of "3", but it can readily happen with other printed forms and with all script forms that I can remember seeing.

Altogether Knobel suggests six changes in a numerical value, and our unknown suggests another. In addition, Knobel suggests three changes in the name of the star used. This makes a total of ten suggested changes in nine readings, with changes being suggested in both elements of the first entry for 1002 March 1. We cannot tell which reading is correct for 1004 January 24 because the indicated times are too close together; this leaves nine readings for which we can make a choice.

Of these nine, Knobel's suggestions prove to be correct twice. The printed form (Caussin) is correct three times; two of these occur for a single entry, namely the first entry for 1002 March 1. Neither form is correct for four of the nine entries.

To the extent that we can apply these results to other recording errors that we find in the ancient and medieval sources, this suggests that simple confusion between similar forms may not be the most common reason for recording errors. More likely, it seems, the recorder or his copyist put down something that happened to be on his mind rather than what he intended to put down. I feel sure that this is the source of many of the errors that occur in the rough drafts of my work, and I doubt that many of them come from my misreading of my handwritten notes or records of computation.

The results that puzzle me are the ones in which Caussin's readings prove correct. I do not see why Knobel was moved to suggest changes in these cases, since there was no problem with Caussin's readings.

CHAPTER X

ANALYSIS AND DISCUSSION

1. The Parameter y as a Function of Time

We can now bring together all the estimates of y that
have been formed from the data presented in this volume.
These estimates are summarized in Table X.1. The estimates
that have been formed from the data presented in volume 1,
modified slightly as a result of introducing the DE102
ephemeris, are summarized in Table A.I.18 in Appendix I.

In preparing Table X.1, I have grouped the data in two
different ways. First, I have grouped together the data
that come from a particular type of observation. For example, the measurements of the interval between moonrise or
moonset and sunrise or sunset are considered separately from
all other kinds of measurements. Then, within each type of
observation, I have grouped the data in accordance with the
time spans that were used in volume 1 and that are tacitly
defined by Tables A.I.1 to A.I.17 in Appendix I.

There are two exceptions to the grouping by time span.
First, there are so few measurements of the times of lunar
eclipses made by Babylonian astronomers (Table VIII.3) that
I have put them together even though their dates range from
-522 to -211. Second, the observation of the solar eclipse
of -321 September 26 has such unusual properties that I have
considered it separately.

In Table X.1, the first column gives the average year
of the observations within a particular group. The second
column gives the estimate of y formed from the data in that
group, and the third column gives the estimate $\sigma(y)$ of the
standard deviation of y. The values of $\sigma(y)$ have been
formed in various ways that have been explained as each
value of y was derived. The fourth column of Table X.1
gives the section in this volume in which the estimate was
made, and the fifth column gives the type of observation in
question.

For the entries dated -373 and -321, the standard deviation is given in the form "+a/-b" rather than as a single
number. For these entries, the error distribution is not
symmetrical. For example, for -373, there is about one
chance out of three that y lies between -22.7 and -22.3.
There is the same chance that it lies between -22.7 and
-24.1. This lack of symmetry comes from the fact that the
data for each entry include an eclipse that happened close
to sunset or sunrise, as the case may be. Thus, in one
direction, the uncertainty in y is set by this circumstance
while, in the other direction, it is set by the expected
errors in measuring time.

Now we have two tables that give estimates of y as a
function of time. Table A.I.18 gives y as a function of
time on the basis of the "amateur" observations of solar
eclipses, while Table X.1 gives y as a function of time on
the basis of all other types of observation. We are now

ready to combine the two tables and to give y̱ as a function of time on the basis of all the available observations that involve the moon. The results are shown in Table X.2. In this table, the first column gives the average epoch† of the

TABLE X.1

A SUMMARY OF THE ESTIMATES OF y̱ FROM VOLUME 2

Epoch	y̱	σ(y̱)	Section	Type of observation
- 567	-10.5	5.8	App. V	Moonrise and moonset
- 567	-22.6	6.4	App. V	Lunar conjunctions
- 441	-38.3	7.8	VIII.2	Moonrise and moonset
- 378	-22.9	13.3	App. V	Moonrise and moonset
- 378	-25.5	3.6	App. V	Lunar conjunctions
- 373	-22.7	+0.4/-1.4	VIII.4	Times of lunar eclipses
- 321	-22.1	+0.9/-2.3	VIII.5	Time of solar eclipse
- 252	-29.0	6.0	App. V	Moonrise and moonset
- 252	-19.1	1.3	App. V	Lunar conjunctions
- 250	-25.1	5.2	VIII.2	Moonrise and moonset
- 250	-20.3	2.8	VIII.3	Lunar conjunctions
- 135	-22.8	3.3	VIII.5	Time of solar eclipse
- 88	-24.8	2.7	VIII.6	Time of solar eclipse
364	-28.4	5.0	VIII.7	Times of solar eclipse
506	-20.0	4.6	VIII.7	Lunar conjunctions
622	-15.7	6.3	V.6	Mean lunar elongation
932	-19.8	2.8	IX.3	Magnitudes of solar eclipses
941	-16.5	0.8	IX.4	Times of solar eclipses
948	-19.7	0.9	IX.2	Times of lunar eclipses
979	-18.8	2.4	IX.1	Lunar eclipse at moonrise
1000	-34.9	18.3	V.2	Mean solar longitude
1000	-19.3	9.2	V.2	Mean lunar elongation
1092	- 5.4	11.7	IV.6	Time of lunar eclipse
1221	- 1.4	25.0	IV.5	Magnitude of solar eclipse
1260	-46.9	40.0	IV.4	Mean lunar elongation
1333	-30.9	16.3	IV.2	Mean lunar elongation
1336	+29.1	21.5	IV.3	Measured lunar longitude
1472	-23.2	7.9	III.4	Times of lunar eclipses
1480	-24.2	7.8	III.5	Times of solar eclipses

†If the observations within a group have greatly different standard deviations, the epoch in Table X.2 is the weighted average of the epochs of the individual observations.

relevant data, the second column gives the estimated value of y at that epoch, and the third column gives the standard deviation $\sigma(y)$ to be attached to y. I continue to make exceptional entries for -373 and -321.

The entry for the year 1850 has a different basis from the other entries. The other entries are all based upon observations that involve the moon in some way. The entry for 1850 is derived from the observations of the sun that Spencer Jones [1939] gave in his famous paper. This entry is obviously independent of the value used for $\dot{n}_M$, the acceleration of the moon, while the other entries do depend upon $\dot{n}_M$ in the way that has been indicated.

TABLE X.2

VALUES OF y OBTAINED BY COMBINING VOLUMES 1 AND 2

Epoch	y	$\sigma(y)$
- 660	-24.4	2.3
- 556	-18.8	2.5
- 394	-24.8	3.1
- 373	-22.7	+0.4/-1.4
- 321	-22.1	+0.9/-2.3
- 251	-19.9	1.1
- 161	-22.6	1.3
- 60	-20.7	1.9
122	-22.0	3.1
394	-24.2	3.2
569	-18.7	2.9
772	-38.1	7.9
878	-21.0	6.4
947	-18.1	0.6
1128	-10.5	4.3
1174	+ 4.0	7.8
1247	-13.6	5.2
1341	-12.0	10.5
1479	-22.0	5.3
1850	- 9.1	2.8

The values from Table X.2 are plotted in Figure X.1 in Section X.2. A circle in the figure is a value of y from the second column of the table, while the bars extending up and down from the circle indicate the standard deviations. The bars are symmetrical about the circles except for the epochs -373 and -321. The error bars for these years are unsymmetrical, for the reasons that have been explained.

The curve in Figure X.1 is the quadratic that best fits the values of y in a weighted least-squares sense. However, the existence of the values with unsymmetrical error bars complicates the finding of this quadratic, because we do not know a priori which of the two possible weights to assign to the values for -373 and -321. In order to resolve this point, I first derived a quadratic using all the values except those for -373 and -321, and I then evaluated the quadratic at the years -373 and -321. For both years, the quadratic derived this way is algebraically greater than the value in the table. This means that the values for -373 and -321 both affect the curve only through their "positive" standard deviations. Therefore, I repeated the derivation of the quadratic using 0.4 as the standard deviation for -373 and 0.9 as the standard deviation for -321.

The curve shown in the figure is the one derived this way. The derivation of the quadratic is facilitated if we take the time origin to be approximately at the center of the data. As I did in volume 1, page 447, and in Section I.5 of this volume, I take the time origin to be the year 600. If we let C denote time in centuries measured from 600, the best-fitting quadratic is

$$y = -19.86 \pm 0.83 + (0.487 \pm 0.102)C$$
$$+ (0.0229 \pm 0.0158)C^2. \tag{X.1}$$

Equation I.19 of this volume gives the value of y we obtain from the data in volume 1, but referred to the epoch with Julian day number 2 375 000.5. This is a day in the year 1790. If we transfer the origin to the year 600, Equation I.19 becomes

$$y = -18.02 + 0.5255C + 0.01515C^2, \tag{X.1a}$$

without the standard deviations. The eclipse and the non-eclipse data agree well with each other.

The non-eclipse data also supplement the eclipse data nicely in several places. The non-eclipse data in particular provide tight limits to y during the Babylonian period from -373 to -161 and the Islamic period centered on 947. They have also reduced $\sigma(y)$ from 17.9 to 10.5 in the 14th century and from 30.0 to 5.3 in the 15th century.

The coefficient of C^2 is interesting. The coefficient of C^2 in Equation X.1a is just about equal to its standard deviation and so it has only marginal significance. The coefficient in Equation X.1 is equal to $1\frac{1}{2}$ standard deviations and thus it has fair but not overwhelming significance. Of course, we know that y cannot have been a linear function of time throughout geologic time, and that it must have higher-order terms that limit it. Equation X.1 tells us with fair assurance that y had a maximum† within historic

---

†Of course Equation X.1 shows an algebraic minimum of y. In saying that it had a maximum, I am speaking of its size.

times and that it has been decreasing in the recent past.
In fact, according to Equation X.1, the maximum of $|y|$ came
in the year -463, and the value of y at that time was -22.4.

Let us look for a moment at the fitting process by
which we found Equation X.1. At a time t_i from Table X.2,
we have a value y_i and an associated standard deviation,
σ_i, say. When we find the best-fitting quadratic function
of time, we start by defining a quantity R_i for each value
of t_i:

$$R_i = (a/\sigma_i) + b(t_i/\sigma_i) + c(t_i^2/\sigma_i) - (y_i/\sigma_i).$$

We then find the values of a, b, and c that make the sum of
the R_i^2 a minimum. When we have found a, b, and c, we calculate the R_i and find their standard deviation. If the
values of σ_i have been chosen in a valid manner, and if a
quadratic is an adequate representation of y, the standard
deviation $\sigma(R_i)$ of the R_i should be unity. In this case,
$\sigma(R_i)$ equals 1.32, which is not far from unity. Thus the
individual σ_i have generally been chosen in a valid fashion,
and a quadratic is capable of representing the time dependence of y with reasonable accuracy.

At the year 600, when $C = 0$ in Equation X.1, the value
of y is 19.86 ± 0.83. That is, we know y in the year 600
with a standard deviation that is less than 1 part in 10^9
per century. We can see by referring to Table X.2 or Figure
X.1 that we also know y in the Babylonian period around -350
and in the Islamic period around 950 with an accuracy that
is also better than 1 part in 10^9 per century. I think we
may fairly say that we know y within 1 part in 10^9 per century for any time between about -600 and +1200 or so. Surprisingly, we know it with less accuracy, say with a standard deviation of about 2, in the modern period.

TABLE X.3

A SUMMARY OF THE ESTIMATES OF $\dot{n}_M$

Epoch	$\dot{n}_M$	$\sigma(\dot{n}_M)$	Section
-522	-22.2	4.9	VIII.4
926	-26.6	8.6	IX.1

2. Estimates of the Lunar Acceleration $\dot{n}_M$

The magnitude of a lunar eclipse obviously does not depend upon the position of the observer, except possibly for very small effects that need not concern us here. This means that the magnitude is independent of the angular position of the earth upon its axis, and therefore the magnitude is independent of the spin acceleration y. However, the magnitude of a lunar eclipse does depend upon the acceleration $\dot{n}_M$ of the moon in its orbit.

In this volume, we have been able to make two estimates of $\dot{n}_M$, which are summarized in Table X.3. In the table, the first column gives the epoch of the estimate, the second column gives the value of $\dot{n}_M$, and the third column gives the standard deviation to be attached to this estimate. The last column gives the section of this volume in which the estimate is formed.

Both estimates agree reasonably well with the value -28 that I have adopted as the standard value in this work, so there is no evidence to suggest that $\dot{n}_M$ has varied appreciably with time during the historical period. If it has varied during the historical period, its variation has almost surely been by much less than a factor of 2. In contrast, y has varied by a factor of more than 2.

The value of $\dot{n}_M$ that I actually derived in volume 1 is (volume 1, p. 26) -28.4 ± 5.7; this was based upon data from about 1650 to 1950. If we assume that $\dot{n}_M$ has been constant since -522, we can combine this estimate with the estimates in Table X.3 and get

$$\dot{n}_M = -25.1 \pm 3.4. \qquad (X.2)$$

In spite of this, I will not change the standard value of -28 that I have been using in the analysis. My main reason for this is that Williams, Sinclair, and Yoder [1978] summarize seven determinations of $\dot{n}_M$ ranging from that of Spencer Jones in 1939 to their own in 1978. The weighted mean that I find from all these determinations is -26.6 ± 1.4.† If we combine this with the value in Equation X.2, we get -26.4 ± 1.3. This does not differ enough from -28 to warrant revising the analysis.

---

†It may not be valid to combine these determinations, because some of them have an atomic time base and some an ephemeris time base. The standard deviation of 1.4 is a purely formal one and is, I suspect, somewhat too small.

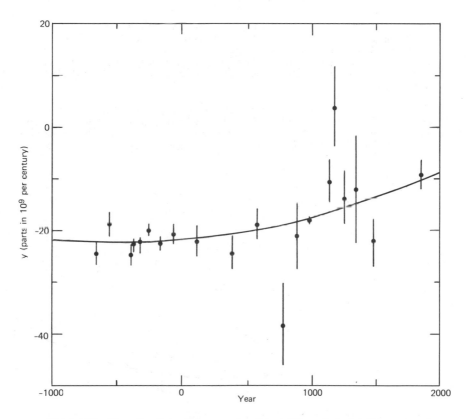

Figure X.1. The spin parameter y as a function of time. A circle with an error bar through it gives the weighted average of y as determined from all observations within a time span centered at the circle, and the half-length of the error bar is the standard deviation of the average. The curve is the quadratic that best fits the plotted points. Note that the error bars are not symmetrical about the central value for the points at -373 and -321.

3.   "Population Bias" in the Observation of Solar Eclipses

In this section, I am reverting to the observations of solar eclipses that were used in volume 1. Those observations, we remember, are purely qualitative ones; they are records such as those typically found in municipal or monastic annals, which simply say that a solar eclipse was observed on a particular day, sometimes with a statement that the eclipse was large or total, and sometimes with a vague estimate of the magnitude of the eclipse for eclipses that were definitely partial.

In my use of such records, I have done something that is diametrically opposed to all other recent studies. I have used all reliable records of eclipses, whereas all other recent students of the subject that I know of have used only the records which indicate a large or total eclipse.

Muller and Stephenson [1975], in particular, have objected to my procedure, with the claim that using all the data introduces a bias. They write [p. 472]: "We consider a statistical approach to large solar eclipses in medieval

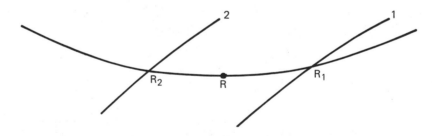

Figure X.2. The non-existence of a "population bias" in the records of solar eclipses. Here we assume that all observations were made at the same point R. The eclipse path is just as likely to pass to the west of R as to the east, so there can be no bias in inferring the spin acceleration even in this extreme case.

times as of doubtful validity on account of the non-random
distribution of medieval centres of population." By "a
statistical approach", they mean the approach of using all
the data, and they use the term "population bias" to mean
the alleged effect coming from the fact that the sources of
medieval annals are not distributed uniformly in longitude
and latitude over the whole earth, but are concentrated in a
fairly small part of the earth's surface.

I showed in Section III.3 of volume 1 that such a bias
is a physical impossibility. There would not be a bias even
if all the records used came from a single location, provided that all the records from that site are used. The
danger comes from using only a part of the records from a
particular location, not from using all of them.

Unfortunately, Lambeck [1980, p. 314] has raised the
question again since the appearance of volume 1, and this
makes it desirable to comment on the matter again here.
Lambeck writes that Muller and Stephenson "argue convinc-
ingly . . . that the population bias . . . is probably responsible" for the "aberrant results" in my earlier work,
particularly in my study of medieval annals [Newton, 1972].

Lambeck words the argument thus: " . . . the use of
partial eclipses may lead to serious population biases† in
that a least-squares estimation process tends to place the
narrow central path through the middle of the observatories,
whereas the real path may be considerably displaced from
this. This would not be very serious if the observers were
quite uniformly distributed along both sides of the path of
totality but, with the tendency for monasteries and towns
during the Middle Ages to cluster in certain areas of
Europe, this condition is seldom met."

This argument confuses the residual of an observation
with a bias in the observation. To see this, let us go to
the extreme case and assume that all the observers were at
the same place, say Rome. Let us represent Rome by the
point R in Figure X.2. The circular arc through R represents the parallel of latitude that passes through Rome. In
deducing a value of y from the fact that an eclipse was seen
in Rome, what matters is where the eclipse path crosses this
parallel of latitude.

The eclipse path marked "1" crosses the parallel to the
east of Rome. The analysis of this eclipse to find a value
of y would advance the rotation of the earth until R is
brought to the point R_1. This would yield a value of y that
is too large. That is, the residual of y for this particular observation would be positive, but this does not constitute a bias. The sun and moon in their orbits do not
know where Rome is, and they are just as likely to put the
crossing point to the west of Rome as to the east. Some
other eclipse will have the path marked "2", and the residual for this eclipse will be negative. Thus, when we use

―――――――――――――

†The emphasis is in the original.

all the observations, both of partial and of total eclipses, the average value of $\bar{y}$ will approach the correct value in the usual statistical fashion. There is no bias in $\bar{y}$ produced by the non-random distribution of the observers over the surface of the earth.

It is probably worthwhile to repeat another important point: Records which specifically state that an eclipse is total do not have the unique value which Muller, Stephenson, and most other workers in the field have given them. I showed this in volume 1, Table XIV.2, page 458. There I examined a quantity μ for an observation of a solar eclipse. If the magnitude of the eclipse is $\underline{m}$, μ denotes $1 - \underline{m}$; μ is 0 for a total eclipse. In preparing Table XIV.2 of volume 1, I studied the actual values of μ for the eclipses that were explicitly reported as total. For such eclipse records from parts of the world other than China, the standard deviation of μ is 0.030, and for the Chinese records the standard deviation is 0.058. That is, if a Chinese record explicitly says that an eclipse is total, the magnitude was actually less than 0.942 about a third of the time. The corresponding number for other parts of the world is 0.970. Thus records which explicitly say that an eclipse was total merely put the observer somewhere within a rather wide zone. They do not put the observer within the narrow zone of totality.†

When he refers to my "aberrant" results, which he attributes to the "population bias" in my data sample, Lambeck [p. 316] seems to mean two different things. One is the variation of $\bar{y}$ (or of $\bar{D}''$ in my earlier work) with time, which I have found in every study of the accelerations that I have made. However, this is a genuine result and not an aberration.‡

. Another "aberrant" result is the unusually large value of $\underline{\dot{n}}_M$ that I found in my first two studies of the accelerations [Newton, 1970 and 1972]. It is almost sure that the value I found there is not correct, and there are two main reasons for the error. First, I had not realized in those early works that Ptolemy's observations were fabricated, and I used them as genuine. Second, I did not appreciate how poorly the ancient and medieval lunar data determine the pair of accelerations $\underline{\dot{n}}_M$ and $\underline{\bar{y}}$. This means that small errors in the data make large errors in the individual accelerations. These considerations, and not a bias in the

---

†Incidentally, the records of "total" eclipses contribute about a third of the total statistical weight of the data.

‡The existence of a time variation is, in fact, shown by the results of Muller and Stephenson, and by all other relevant writers that I know of. The other writers have simply overlooked this variation in their discussions of their results, but the variation itself is there in their data, and it is not a peculiarity of my analyses. See page 59 of volume 1.

location of ancient observatories, are the reasons for the aberrant value of $\underline{n}_M$.

4.   Instantaneous and Average Values

Equation X.1 gives the value of $\underline{y}$ as a function of time, in which the origin of the time $\overline{C}$ has been taken as the year 600 for convenience in the fitting process. However, the values of $\underline{y}$ in Table X.2 are actually referred to the epoch with the Julian day number 2 375 000.5, which is sometime in the year 1790. Now we need to have the functional form of $\underline{y}$ referred to 1790 as well. If we let $\underline{t}$ denote time in centuries from 1790, $\underline{y}$ becomes

$$\underline{y} = -10.82 + 1.032\underline{t} + 0.0229\underline{t}^2. \qquad (X.3)$$

There is no simple way to derive the standard deviations of the coefficients in Equation X.3 from those in Equation X.1, so I will not try to derive the standard deviations in Equation X.3.

The fact that $\underline{y}$ varies with time forces us to look more carefully at how it is defined. For the moment, let us revert to using the angular acceleration $\dot{\omega}_e$ of the earth's spin rather than the parameter $\underline{y}$. In the calculations that lead to Table X.1 or X.2, I take the angular displacement of the earth around its axis† to be $\tfrac{1}{2} \dot{\omega}_e \underline{t}^2$. This means that $\dot{\omega}_e$ is not the actual or instantaneous acceleration at any time. Instead, it is some kind of average acceleration between 1790 and the time in question. Since $\tfrac{1}{2} \dot{\omega}_e \underline{t}^2$ is the displacement, the instantaneous acceleration is the second derivative of this quantity.

Although we customarily use Ω to denote angular momentum rather than angular velocity, let us be unconventional for the moment and use $\dot{\Omega}_e$ to denote the instantaneous angular acceleration of the earth. Then

$$\dot{\Omega}_e = (\underline{d}^2/\underline{d}\underline{t}^2)(\tfrac{1}{2} \dot{\omega}_e \underline{t}^2).$$

Conversely,

$$\dot{\omega}_e = (2/\underline{t}^2) \int\!\int \dot{\Omega}_e \, dt^2.$$

That is, $\dot{\omega}_e$ is an average value that I will call, for want of a standard term, the "time-squared" average of $\dot{\Omega}_e$.

Obviously, we can define a time-squared average for any time-dependent quantity. If we let $\underline{Y} = 10^9 (\dot{\Omega}_e / \omega_e)$, in

---

†By this, I mean the deviation of the angular position from a uniform rotation.

parallel with the definition that $\underline{y} = 10^9(\dot{\omega}_e/\omega_e)$, we see that $\underline{y}$ is the time-squared average of Y. For $\bar{a}$ reason that will appear in a moment, we also want $\overline{\text{the}}$ time-squared average of M^2, in which $\underline{M}$ is the magnetic dipole moment of the earth. I will use $\overline{M^2}$ to denote this average.

Smith [1967] has estimated M at about the year -1050, at four epochs between about 300 and 1600, and at the present, which I take to be 1960. The squares of his values, along with the associated error bars, are plotted in Figure X.3; the error bar at 1960 is too small to plot on the scale of the figure. In order to find $\overline{M^2}$, we will need a convenient mathematical representation of the values of M^2.

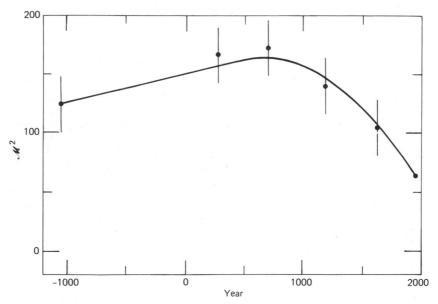

Figure X.3. The square of the earth's magnetic dipole moment as a function of time. The (reduced) dipole moment $\mathscr{M}$ is expressed in units of 10^{25} gauss-cm^3, so the point marked 200 on the vertical axis, for example, corresponds to $\mathscr{M}^2 = 200 \times 10^{50}$ gauss^2-cm^6, or to $\mathscr{M} = 14.14 \times 10^{25}$ gauss-cm^3. The solid circles show the values of $\mathscr{M}^2$ at different epochs, and the error bars show the standard deviations. The standard deviation at the epoch 1960 is too small to plot. The curve is made up of a quadratic and a straight line tangent to it, and it is used to represent $\mathscr{M}^2$ as a function of time in the analysis.

The values in Figure X.3 can be represented well by the combination of a parabola and a straight line. If we continue to let t denote time in centuries from 1790, these functions are

$$M^2 = 193.3 + 2.407t, \qquad t < -13.25,$$
$$M^2 = 86.95 - 13.65t - 0.6058t^2, \qquad t > -13.25.$$

The time $t = -13.25$, which is the year 465, is the point where the straight line is tangent to the quadratic. I will explain how the quadratic was chosen in Section X.6. Once the quadratic was chosen, the straight line was chosen to pass through the point at -1050 and to be tangent to the quadratic.

When we do the integrations needed to find the time-squared average $\overline{M^2}$, we get

$$\overline{M^2} = (3113/t^2) + (939.8/t) + 193.3$$
$$+ 0.8025t, \qquad t < -13.25,$$

$$\text{(X.4)}$$

$$\overline{M^2} = 86.95 - 4.550t - 0.100\ 97t^2, \quad t > -13.25.$$

5. The Sources of the Accelerations

When I chose The Moon's Acceleration and Its Physical Origins as the title of this work several years ago, I was unhappy with the possible ambiguity in it. I was afraid that the reader might think that the physical origins in question were the origins of the moon rather than the origins of its acceleration. However, I could not think of an unambiguous alternative that was short enough to fit comfortably onto the spine of a book, so I left the title as it stands. I believe that most readers have realized what I meant; at any rate, I have heard no objections.

At the time I chose the title, I was using mean solar time rather than dynamical time as the time base, and by the acceleration of the moon I meant its acceleration with respect to solar time throughout volume 1. In this volume, for reasons that were explained in Chapter I, I have changed to dynamical time. When I speak of the sources of the accelerations, most of the discussion will center on the acceleration of the earth's spin with respect to dynamical time. However, the reader should remember that the acceleration ν_M' of the moon with respect to solar time equals

$$\nu_M' = \dot{n}_M - 1.7373y. \qquad \text{(X.5)}$$

Thus any source of the acceleration y is also a source of the moon's acceleration with respect to solar time, and it is not a paradox to deal mostly with the earth's spin acceleration in a book whose title deals with the moon's orbital acceleration.

The only source of $\dot{n}_M$ that is large enough to matter in this discussion is friction in the lunar tide, so far as we know. This friction acts to take angular momentum out of the earth's spin and to put it into the moon's orbital motion. The only way the moon can increase its angular momentum is to move into a larger orbit, in which it has a smaller angular velocity. Thus friction in the lunar tide gives rise to negative values of both $\dot{n}_M$ and y.

In the discussion of the sources of the earth's spin acceleration in volume 1 (Section XIV.9), I based the discussion on the instantaneous value Y rather than the time-squared average y. Since only one source varies significantly with time, so far as we know, I believe it is easier to talk about y, and I will do so here.

There are many physical effects that can give an acceleration to the earth's spin, but at most five of all those that have been suggested can be large enough to matter at the present level of accuracy. I will take up these five in turn, and I will designate their estimated contributions to y by y_1 through y_5.

1. The earth's magnetic field. It is conceivable that the earth's magnetic field, which mostly originates in the core, acts on magnetic material in the mantle or crust and interchanges angular momentum between the core and, say, the mantle. This would be interpreted by observers on the surface, which is tied to the mantle, as an angular acceleration of the earth. This contribution to y, which I call y_1, could be proportional either to M, the earth's magnetic dipole moment, or to M^2. In volume 1 (pages 474ff), I gave two arguments which both indicate that a proportionality to M^2 is more plausible, and I consider only this possibility here.

Of all the possible sources of y, only the magnetic field is known to have changed by an important amount within historic times. This suggests trying to fit y to a constant plus a multiple of $\overline{M^2}$, which is the time-squared average of M^2 that was derived in the preceding section. The function of this form that best fits the data is $y = 8.9 - 0.2268\overline{M^2}$ The term proportional to $\overline{M^2}$ is the magnetic contribution y_1 and the constant term is the total contribution from all other sources. That is,

$$y_1 = -0.2268\overline{M^2} \qquad (X.6)$$
$$y_2 + y_3 + y_4 + y_5 = +8.9. \qquad (X.7)$$

At the moment, I will say only that these values agree excellently with the values of y in Table X.2 and Figure X.1. I will discuss the agreement in more detail in Section X.6.

A correlation between $\overline{M^2}$ and $\underline{y}$, if it is indeed real, does not necessarily imply that the core and mantle interact only through the mechanism of the magnetic field. So far as we know, it is possible for the core and mantle to interact directly at their interface. Such an interaction might affect the properties of the core and hence the properties of the magnetic field. In other words, a direct interaction at the interface might produce a correlation of the kind we have found. I will return to this question in Section X.7.

2. Tidal friction. The friction in the lunar tide has already been discussed qualitatively. There is also a solar tide which contributes to $\underline{y}$, and we lump the two tides together for convenience. Further, there are theoretical connections between friction in the two tides, so we can relate the total friction in the two tides to the angular acceleration $\underline{\dot{n}}_M$. The result is (volume 1, page 471) that the tidal contribution $\underline{y}_2$ equals $1.165\underline{\dot{n}}_M$; the coefficient, which is calculated from various geophysical quantities, is uncertain by about 0.03. Hence, if $\underline{\dot{n}}_M = -28$,

$$\underline{y}_2 = -32.6. \qquad\qquad\qquad (X.8)$$

Several recent writers, including Jeffreys [1982] and Lambeck [1980], have questioned the need for any contribution to $\underline{y}$ other than the tidal one. Lambeck [p. 289] words the matter this way: "However, evidence for the non-tidal acceleration remains weak due to it being a relatively small difference between two larger quantities that are both known to within about 10% only." This ignores several important factors in the situation:

a. The non-tidal contribution is not a small difference between two larger quantities. If $\underline{\dot{n}}_M = -28$, the tidal contribution to $\underline{y}$ is (Equation X.8) -32.6 while $\underline{y}$ itself in the year 1790 was (Equation X.3) -10.8. The nontidal contribution, which is the difference, is $+21.8$. This is twice the size of $\underline{y}$.

b. If we assume with Jeffreys and Lambeck that $\underline{\dot{n}}_M$ has been constant within the historical period, we must conclude that the tidal contribution to $\underline{y}$ has also been constant. Since $\underline{y}$ is a strong function of time, there must be a large non-tidal contribution to supply the necessary time dependence.

c. The uncertainties in $\underline{\dot{n}}_M$ and $\underline{y}$ are actually greater than the 10 per cent that Lambeck mentions; they are closer to 20 per cent. Nonetheless, this does not produce much uncertainty in the value of the non-tidal contribution. The argument assumes that the uncertainties in $\underline{\dot{n}}_M$ and $\underline{y}$ are uncorrelated when in fact they are highly correlated by the ancient and medieval data. The uncertainty in the non-tidal contribution is much smaller than the uncertainty in either $\underline{\dot{n}}_M$ or $\underline{y}$.

TABLE X.4

THE NON-TIDAL COMPONENT OF THE SPIN ACCELERATION[a]

$\overset{\bullet}{\underline{n}}_M$	$\underline{y}$[b]	$\underline{y}(tidal)$[b]	$\underline{y}(non\text{-}tidal)$[b]
-28	-19.9 ± 0.8	-32.6 ± 0.8	+12.7 ± 1.1
-22	-14.9 ± 0.8	-25.6 ± 0.7	+10.7 ± 1.1

[a]For the year +600.

[b]In parts in 10^9 per century.

To see the latter point, look at Table X.4. For the year 600 (Equation X.1), the value of $\underline{y}$ is -19.9 ± 0.8 if the value of $\overset{\bullet}{\underline{n}}_M$ is -28. The tidal contribution is $(1.165 \pm 0.03)\overset{\bullet}{\underline{n}}_M$, which is -32.6 ± 0.8. The uncertainty in this value comes from the coefficient of $\overset{\bullet}{\underline{n}}_M$ and is independent of the uncertainty in $\underline{y}$. The non-tidal contribution is then +12.7 ± 1.1, so that the estimate of the non-tidal contribution is more than ten times its standard deviation.

If we change $\overset{\bullet}{\underline{n}}_M$ to -22, say, we must first decide what value of $\underline{y}$ we would deduce from the data. We must remember that the ancient and medieval data determine the value of the lunar elongation at a specific epoch. If we change $\overset{\bullet}{\underline{n}}_M$ to -22 without changing $\underline{y}$, we add $\frac{1}{2} \times 6 \times (6.00 - 19.69)^2 = 562''$ to the longitude of the moon in the year 600 without changing the longitude of the sun. Hence we add 562'' to the elongation. We must decrease our estimate of ΔT (ephemeris time minus solar time) by enough to remove this change in the elongation. Since the elongation changes by 0.508 seconds of arc per second of time, we must subtract 1 107 seconds from ΔT and hence we must add (Equation I.15) 5.0 to $\underline{y}$. This makes $\underline{y}$ = -14.9, as Table X.4 shows, without changing its standard deviation appreciably.

At the same time, as the table shows, we change the tidal contribution to -25.6 ± 0.7, and thus we change the estimate of the non-tidal contribution to +10.7 ± 1.1. The non-tidal contribution changes by only a third of the change in $\overset{\bullet}{\underline{n}}_M$.

3. The moment of inertia of the solid earth and core. If the core is still growing, the result is to take material from the solid upper parts of the earth and to put it into the liquid core where it has a smaller radius. This causes a decrease in the moment of inertia of the earth and hence an increase in its angular velocity. In other words, this process gives rise to a positive value of the contribution to $\underline{y}$ that we are calling $\underline{y}_3$.

Lyttleton [1982, p. 81] estimates, from a study of ancient solar eclipses, that this contribution is 1.40 in a system of units in which the tidal contribution y_2 is -5.99.† If we change to our system of units in which y_2 = -32.6, this gives y_3 = +7.6.

Unfortunately, it is logically impossible to derive a value of y_3 from a study of ancient eclipses. The information we get from a study of eclipses (or from any of the data used in this work) is purely kinematic, and it can tell us nothing about the physical origins of the acceleration. It can tell us the total value y, but not how to partition it among various physical contributions. What Lyttleton does is to derive a value of y from a study of ancient eclipses, ignoring all other data. He then postulates that y = y_2 + y_3, while saying that y_1, y_4, and y_5 are all zero.

Now, it is certain that y_1 is an important contribution, at least if we use y_1 to mean simply a contribution with the required time dependence. Lyttleton ignores this important contribution. Since he omits an important feature of the situation from his analysis, his value of y_3 is not physically significant. We must attempt to derive y_3 from more fundamental principles. In order not to take us too far afield here, I present the necessary derivation in Appendix VII.

As I show in the appendix, if the radius of the core increases by 1 meter, the moment of inertia of the earth decreases by 1.081×10^{31} kilogram-meters^2. The only parameters that enter into this calculation are the radius of the core, the radius of the earth, and the densities of the core and mantle at their interface, so this figure should be rather accurate. Since the moment of inertia of the earth is about 8×10^{37}, the change is -135.1 parts in 10^9 per meter. Hence the spin accelerates by 135.1 parts in 10^9 per meter's growth of the core.

Lyttleton [p. 100] estimates that the core grew from nothing to a radius of about 2 042 kilometers in a time that is negligible on the geological time scale, and he further estimates that this happened at a time somewhere between 2.5 and 4.0 billion years ago. If we use the shorter time span, the radius has been growing 0.0572 meters per century, on the average. If we assume that this rate prevails today, we get

$$y_3 = +7.7. \qquad\qquad (X.9)$$

This is close to the value Lyttleton gets, but this agreement is a coincidence.

There is another effect of the solid part of the earth that is negligible at the present level of accuracy. The

---

†By changing some of the geophysical parameters involved, he also gets y_3 = 1.15 and y_2 = -5.74. For simplicity, I will use only the larger values.

earth's shape is close to the shape it would have if it were in hydrostatic equilibrium under the influence of its rotation and its self-gravitation. As its rate of rotation decreases, it becomes less flat. This means that its equatorial radius and its polar moment of inertia both decrease. This gives a contribution to y.

The matter is complicated by the fact that the present flattening is not the equilibrium value [Henriksen, 1960]. Instead, it is the flattening that corresponds to the rate of rotation some tens of millions of years ago. By making some plausible assumptions about the way the actual moment of inertia differs from the hydrostatic one, I estimate that the contribution of this effect to y is of the order of 0.01. We can neglect this at the present level of accuracy.

The composition of the liquid core, and the way in which it grows, are matters on which there is disagreement among geophysicists. It is not my intention to take sides in this discussion. I have used Lyttleton's values relating to the core only for the sake of illustration.

4. A change in the gravitational constant. In a sense, y_4 is simply everything that is left over after we subtract y_1, y_2, and y_3 from y, because I am going to take zero as the best estimate of y_5. This gives

$$y_4 = +33.8. \tag{X.10}$$

The various cosmological theories which admit a value of $\dot{G}$ that is not zero agree that the resulting angular acceleration has the form

$$\dot{\omega}/\omega = K(\dot{G}/G). \tag{X.11}$$

Unfortunately, the theories disagree about the size of K and even about its sign. The only point of agreement is that $|K|$ is either 1 or 2 in all theories I have seen proposed so far. This means that

$$|\dot{G}/G| = 16.9 \quad \text{or} \quad 33.8 \tag{X.12}$$

parts in 10^9 per century. Even the smaller value in Equation X.12 seems somewhat too large to be compatible with current estimates of the age of the universe.

5. The moment of inertia of the oceans. y_3 represents the contribution to y of variations in the moment of inertia of the solid earth and of the core. I will use y_5 to denote the contribution from changes in the moment of inertia of the oceans or, more accurately, of the oceans plus the continental ice. I accidentally omitted this source from the discussion in volume 1, although I included it in my first discussion of the sources [Newton, 1970, p. 290].

Sea level, the average temperature of the air, and the average temperature of the water are almost certainly rising at the present time, and this trend has probably been present, on the average, for some centuries. Over a longer period, Munk and MacDonald [1960, p. 236] cite a study by

F. Zeuner† that I have not seen which concludes that sea
level has dropped by 2 meters during the past 2000 years.
Since I have not seen this study, I should probably not
criticize it, but I cannot help being skeptical. I know of
two places that are now under water but that were dry land
2000 years ago. One is the area around Mont St. Michel; the
mountain was once definitely part of the mainland, but it is
now an island at high tide. The other is Alexandria. There
large parts of the ancient city are now under water.

A rise in temperature increases the moment of inertia
of the oceans in two independent ways. First, it makes the
oceans expand without changing the mass of water in them.
In Appendix VI, I estimate that a change of 1° Centigrade
changes the earth's rotation by 0.000 25 parts in 10^9 because of the expansion. This effect of temperature is negligible.

Second, a rising temperature makes ice melt. The
melted water runs into the oceans, increasing its level and,
more importantly, its mass. Some of this mass shows up in
the equatorial regions and increases the moment of inertia
of the oceans. This increase outweighs the decrease caused
by the disappearance of the ice. In Appendix VI, Equation
A.VI.9, I estimate that the resulting contribution to $\underline{y}$ is
−13 if sea level is rising at the rate of 10 centimeters per
century as a result of melting ice.

Further, I note that this estimate must be too large in
size, because it neglects the reaction of the solid earth to
changing loads. However, let us make the extreme assumption
of assuming that Equation A.VI.9 is correct and that sea
level has been falling at the rate of 10 centimeters per
century, as Zeuner concludes. Then

$$\underline{y}_5 = +13. \tag{X.13}$$

This reduces the estimate of $\underline{y}_4$ (Equation X.10) from +34 to
+21.

However, the size of $\underline{y}_5$ given above is too large for
the assumed rate of change of sea level, for the reason
stated, so $\underline{y}_5$ is probably not an important factor in the
situation. I also suspect that even the sign of $\underline{y}_5$ is uncertain. In the present state of knowledge, I think that
the best estimate of $\underline{y}_5$ is zero.

6. The Best Representation of y

We now have two fitted curves to represent the behavior
of the spin acceleration parameter $\underline{y}$ as a function of
time. One is Equation X.1 (or Equation X.3, which is the
same except for the origin of time); this was obtained by

---

†The reference is to a book by Zeuner entitled Dating the
  Past; an Introduction to Geochronology, Methuen, London, 1946.

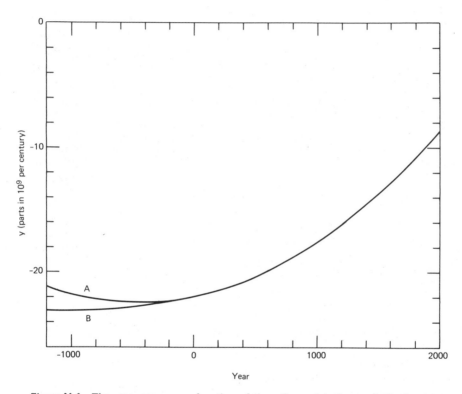

Figure X.4. The parameter y as a function of time. Curve A is the quadratic that best fits the astronomical data, which go back to about –660. Curve B is the function that best fits the data if we assume that y has a component that depends upon the earth's magnetic moment. The curves differ only because the magnetic moment does not follow a quadratic for years before about 0.

fitting a quadratic to the values in Table X.2. The other
is

$$\underline{y} = 8.9 - 0.2268 \overline{M^2},\tag{X.14}$$

in which $\overline{M^2}$ is given by Equation X.4. It is now time to
explain how I found the quadratic portion of Equation X.4.

To do this, I took Equation X.3 for $\underline{y}$, in which the
three coefficients were found by fitting. From this, I constructed the quadratic that represents the instantaneous
value $\underline{Y}$ by the method described in Section X.4. I then
found the constants $\underline{a}$ and $\underline{b}$ which make $\underline{a} + \underline{b}\underline{Y}$ the best fit
to the values of M^2 plotted in Figure X.3. This fit is $M^2 =$
$39.24 - 4.409\underline{Y}$. When we solve this for $\underline{Y}$, we get

$$\underline{Y} = 8.90 - 0.2268 M^2.$$

Equation X.14 then results immediately from this by taking
the time-squared average of both sides.

This description needs one modification. In fitting
the form $\underline{a} + \underline{b}\underline{Y}$ to the values of M^2, I omitted the value at
-1050, for two reasons. First, it is clear from Figure X.3
that a quadratic cannot represent M^2 for times much before
the year 0. Second, we do not have any value of $\underline{Y}$ from the
astronomical data for times before -660, and we would have
to use an extrapolated curve for $\underline{Y}$ in order to use the value
of M^2 at -1050.

If we substitute for $\overline{M^2}$ from Equation X.4 into Equation
X.14, we get

$$\underline{y} = - (706.0/\underline{t}^2) - (213.15/\underline{t})$$
$$- 34.95 - 0.1820\underline{t}, \qquad \underline{t} < - 13.25, \tag{X.15}$$
$$\underline{y} = -10.82 + 1.032\underline{t} + 0.0229\underline{t}^2, \quad \underline{t} > - 13.25.$$

The quadratic part of this is the same as Equation X.3, from
the way it was derived.

In Figure X.4, curve $\underline{A}$ is the plot of Equation X.3
extended back to -1200; we should remember that the earliest
value is for -660. Curve $\underline{B}$ is the plot of Equation X.15.
The curves are identical for $\underline{t} > -13.25$, which is the year
$+465$. On the scale of the figure, though, we cannot see any
difference from $+465$ back to -200. In the year -660, the
time of our earliest value, the difference is about 0.6,
which is probably less than the uncertainty in $\underline{y}$ at that
time. Even in the year -1200, the difference between the
curves is only about 1.9.

The figure shows us that Equation X.14, with a contribution that is related to the earth's magnetic field, is
an excellent representation of the acceleration parameter $\underline{y}$.

It makes no difference which curve we use in Figure X.4 for any time from which we have data in Table X.2. If we need to extrapolate y back to earlier times for some reason, we should definitely use curve B. Therefore, I propose curve B as the best estimate of y we can make on the basis of present knowledge. Of course, we should not use it at times much earlier than -1050 unless we can get an estimate of M for earlier times.

If we change our estimates of the dipole moment M before the present, we will need to change the coefficients in Equations X.14 and X.15, but it seems unlikely that we will need to change them by an important amount. At least, a recent article by Kovacheva [1982] does not indicate the need for any significant changes. However, the article does present some evidence that the magnetic field is varying approximately sinusoidally with a period of about 9 000 years. If this conclusion is confirmed, it will give us a basis for using Equation X.14, but not, of course, Equations X.15, back to about the year -7 000.

7. A Final Summary

In volume 1 of this work, I used a large body of qualitative observations of solar eclipses. By a qualitative observation, I mean one in which the observer noted that a solar eclipse of a particular date was seen at a particular place. He may also have given a qualitative indication of the magnitude of eclipse, such as saying that stars could be seen at the height of the eclipse. Further, he may have given some indication of the time of the eclipse, such as saying that the eclipse ended in late afternoon. However, he did not use instruments to make quantitative measurements of either the time or the magnitude.

In volume 2, I used quantitative astronomical observations that have survived from ancient and medieval times. These include measurements of the time interval between either moonrise or moonset and either sunrise or sunset, the times when the moon was in conjunction with a particular star, the times associated with both lunar and solar eclipses, and the magnitudes of both lunar and solar eclipses. I have also used a few miscellaneous observations, such as the mean elongation of the moon at a certain epoch taken from tables of the sun and moon.

A few of the observations give us estimates of the lunar acceleration $\dot{n}_M$ with respect to dynamical time. When I combine these estimates with other recent determinations, on the assumption that $\dot{n}_M$ has been constant within historic times, I get $\dot{n}_M = -26.0 \pm 2.8$.

In both volumes, although I approached the matter from different directions, I ended up by studying mainly the acceleration y of the earth's spin, on the assumption that the moon's acceleration $\dot{n}_M$ with respect to dynamical time is

-28 $''/cy^2$.† The results obtained from both the qualitative and the quantitative observations agree excellently. The main feature of the situation is that the acceleration $\underline{y}$ is a quadratic function of time within the historic period, and this function is given by Equation X.1. This function gives $\underline{y}$ with an accuracy that is within 1 part in 10^9 per century for any time between about -600 and $+1200$. As we approach the modern period, the accuracy degrades to about 2 parts in 10^9 per century. These accuracies refer to the standard deviation. In extrapolating $\underline{y}$ to earlier times for which we have no data, we should use curve $\underline{B}$ in Figure X.4.

The value of $\underline{y}$ may be biased by the value used for $\underline{\dot{n}}_M$, but it has no appreciable bias resulting from any other cause. The data produce a strong correlation between the values of $\underline{y}$ and $\underline{\dot{n}}_M$, and if we decide to alter the value of $\underline{\dot{n}}_M$ we will also have to alter Equation X.1. However, the main change will probably be in the constant term in $\underline{y}$, with relatively small changes in the linear and quadratic terms in the time.

The data show clearly that there are important contributions to the spin acceleration $\underline{y}$ other than that arising from tidal friction. There is almost surely an important contribution that is proportional to M^2, the square of the earth's magnetic dipole moment. Other contributions are less certain. Lyttleton argues plausibly that the earth's core is still growing at the expense of the solid earth. This phenomenon produces a positive contribution to $\underline{y}$. We have no explanation at the moment for the remaining part of $\underline{y}$ except, possibly, secular changes in the value of the gravitational constant G and in the amount of continental ice (counting Greenland as a continent in this context).

In fact, after we take out the tidal and the magnetic contributions, the remaining part of $\underline{y}$ equals $+41.5$ parts in 10^9 per century. If we ascribe $+7.7$ parts to the growth of the core, and $+13$ to the effects of ice, we still have $+20.8$ parts that we must ascribe to a rate of change of G. We can make little change in this number by any allowable change in the lunar acceleration $\underline{\dot{n}}_M$. Further, since the value $+13$ is known to be an overestimate, the value $+20.8$ must be an underestimate, under our assumptions.

In spite of this, I do not urge these results as strong evidence that $\underline{G}$ is changing with time, for two reasons. For one thing, this conclusion assumes that we have thought of all the important phenomena that contribute significantly to $\underline{y}$. For another, it assumes that we have correctly calculated the other contributions.

We know the tidal phenomena well enough to be rather confident that we have calculated the tidal contribution

---

†The difference between this value and the one just obtained is not enough to warrant revising the analysis.

with an uncertainty of about 1 part in 10^9 per century (for a given value of $\underline{\dot{n}}_M$). We do not know the physics of the magnetic contribution well enough to calculate it from first principles, but the match of the time dependence of $\underline{y}$ to that of M^2 gives this contribution with considerable confidence. The main difficulty comes from deciding what to attribute to the growth of the core and to the effects of ice.

As an illustration, let us suppose that the effect of changing ice is zero. We know rather accurately (Appendix VII) the effect on the rotation rate of adding 1 meter to the core, but we do not know the rate at which the core is growing. Since the earth is still in a state of high tectonic activity, it is plausible that the core is still growing, on a geological time scale.† However, the time scale of the data in this work is only about 2 500 years, and the core may not be growing on this time scale.

That is, the growth of the core may be episodic, with an "episode" that is long compared with 2 500 years. Thus, at the moment (and 2 500 years is a moment in this context), the core may not be growing at all, and $\underline{y}_3$ may be zero. On the other hand, if there are times when the core does not grow, there must be other times when the rate of growth is greater than the average. It is thus possible that $\underline{y}_3$ is several times larger than the average value of +7.7. Certainly there is no difficulty in supposing that the growth of the core and a drop in sea level between them account for all of the needed total of +41.5, leaving nothing to come from a change in $\underline{G}$.

There is another intriguing possibility. If there are episodes in the growth of the core, an episode may cause a change in the magnetic field. It is possible that we are in a phase of the growth that is causing the field to change. The contribution that is proportional to M^2, then, may have its ultimate origin in the growth of the core, so that the core may possibly be the origin of two contributions, those we have called $\underline{y}_1$ and $\underline{y}_3$.

To end on a positive note, the value of $\underline{\dot{n}}_M$ is almost surely in the −20's, and −28 is a plausible value for it. However, I cannot quarrel with an estimate as far removed as −22. The acceleration $\underline{y}$ of the earth's spin is a strong function of time, even within the historical period, and it contains major contributions from sources other than friction in the tides. In fact, some of these other sources may well be larger than tidal friction.

Other work has shown that the power spectrum of the earth's spin contains important components with periods rang-

---

†Note, though, that tectonic activity near the earth's surface does not necessarily imply that the core is still growing.

ing from less than a month to a century or so. The present
work has shown that the power spectrum extends to periods
that must be measured in millenia, and we should be surprised if this were not so.

A REVISED ANALYSIS OF THE RECORDS USED IN VOLUME 1

In the first chapter of volume 1, I used a body of observations made with the telescope and pendulum clock to find the parameters $\underline{\dot{n}}_M$ and $\underline{y}$, which are based upon ephemeris (dynamical) time, and from them I derived values of the accelerations ν_M' and ν_S', which are based upon solar (universal) time. The reference epoch used in this study was the beginning of the year 1800.

In later chapters, I studied a large body of records which state that a particular eclipse of the sun was observed at a particular place. I found that these records do not allow us to find two accelerations with useful accuracy. In fact, the only parameter we could find from the eclipse records is $\underline{D}''$, defined as $\nu_M' - \nu_S'$.

Now, with the introduction of the DE102 ephemeris, we can no longer use the parameters ν_M' and ν_S', for reasons that I discussed in Section I.7. Instead, we can use only the parameters $\underline{\dot{n}}_M$, with a reference epoch in 1969, and $\underline{y}$, with a reference epoch in 1790. Thus, if we are to combine the observations used in volume 1 with those to be used in volume 2, we must revise that part of volume 1 that was implicitly based upon the use of solar time. That is, we must make a new analysis of the records of the solar eclipses that were used in volume 1. We do not need to revise the analysis of the modern observations.

Since we can find only one parameter from the solar eclipses, and since we can no longer use $\underline{D}''$, we must decide what parameter to use in the revised analysis. Clearly we must adopt some result from the analysis of the modern records and assume that it applies throughout the period covered by the records of solar eclipses, which begins in the year -719.

As we said in Chapter I, there are physical reasons for believing that $\underline{\dot{n}}_M$, the acceleration of the moon with respect to dynamical time, has been substantially constant within historic times, and for believing that $\underline{y}$ has varied by a considerable amount. Thus, when we revise the analysis of the solar eclipses, we should take $\underline{\dot{n}}_M$ to be constant and equal to the value found from the modern observations. Using this value of $\underline{\dot{n}}_M$, we can then estimate the parameter $\underline{y}$ at various epochs since -719. The value I use for $\underline{\dot{n}}_M$ is

$$\underline{\dot{n}}_M = -28 \quad ''/cy^2. \tag{A.I.1}$$

I base the analysis of the eclipses upon the coordinate that I call η [Newton, 1972, p. 578]. In defining η, we consider a particular place of observation, and we further

consider the sun-moon line at the instant when the eclipse
is a maximum at the place in question. By η, I mean the
distance from the point of observation to the sun-moon line,
measured at the instant of maximum eclipse, in units of the
earth's radius. The coordinate η is taken to be positive if
the point of observation is northward from the sun-moon axis
at the instant of maximum eclipse.

The ith record furnishes an estimate η_i of the coordinate η for the eclipse in question. To this estimate, we
attach a standard deviation $\sigma_{\eta,i}$ that is based upon the
nature of the record. The assignment of $\sigma_{\eta,i}$ is discussed
in detail in Section XIII.1 of volume 1. It has three components, namely: (1) the uncertainty in the magnitude of
the eclipse that is implied by the record, (2) the uncertainty in the point of observation that is implied by the
record, and (3) the probability that I have drawn the correct implications from the record.†

Let η_{obs} denote the value of η that is implied by the
ith record, and let η_o denote the value of η that we calculate for the time and place of this record, using $\underline{\dot{n}}_M = -28$
(Equation A.I.1) and the conventional value

$$y = -20 \qquad \text{parts in } 10^9 \text{ per cy.} \qquad (A.I.2)$$

I am dropping the i subscript for the moment for the sake of
a simple notation. Let the residual $\underline{R}$ for this record be
defined by

$$\underline{R} = \eta_{obs} - \eta_o. \qquad (A.I.3)$$

We base the inference of $\underline{y}$ upon the collection of the $\underline{R}$'s
and upon the derivative of η with respect to $\underline{y}$. To find
this derivative, I calculate η using the values $\underline{\dot{n}}_M = -28$ and
$\underline{y} = -10$, and perform the obvious numerical differentia-
tion.‡ For the sake of those who may wish to vary $\underline{\dot{n}}_M$, I
also calculate η using the values $\underline{\dot{n}}_M = -18$ and $\underline{y} = -20$ and
find the desired derivative.

---

†Many ancient and medieval writers copied from earlier
 writers without giving any indication that they had done
 so. To them, the happening was the important thing; the
 origin of a particular record was unimportant. In using
 old records of solar eclipses, then, we must decide from
 the literary circumstances where a particular record originated. By item (3), I mean the probability that I have
 correctly identified the place or region where a particular
 eclipse was actually observed.

‡For some records, it was necessary to take $\underline{y} = -15$, or even
 $\underline{y} = -19$, for the second value of $\underline{y}$, but the principle
 remains the same.

In Tables A.I.1 through A.I.17†, I tabulate the follow-
ing quantities, for which I have restored the use of $\underline{i}$ as a
subscript:

$$\underline{r}_{\underline{i}} = \underline{R}_{\underline{i}} / \sigma_{n,\underline{i}} ,$$

$$n_{\underline{Y},\underline{i}} = (1000/\sigma_{n,\underline{i}})(\partial n/\partial \underline{y}), \qquad (A.I.4)$$

$$n_{\underline{N},\underline{i}} = (1000/\sigma_{n,\underline{i}})(\partial n/\partial \underline{\dot{n}}_{\underline{M}}).$$

In addition, I tabulate the values of $\sigma_{n,\underline{i}}$. In the tables,
I have divided the records into "time bins" in the same way
that I did in Tables XIII.5 through XIII.21 (pages 416-443)

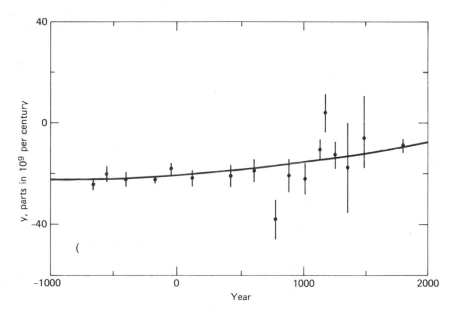

Figure A.I.1. The variation of the acceleration $\underline{y}$ of the earth's rotation with time, as
derived from solar eclipses. The plotted points and associated error bars are those listed
in Table A.I.18 of this volume. The curve is the variation of $\underline{y}$ derived from the results
of volume 1. The agreement is good, although the present results suggest that the
minimum of $\underline{y}$ occurred about the year +200 instead of -1100. The curve is so flat
that the difference in the minimum points is not significant.

---

†The tables are found at the end of this appendix.

of volume 1. I call the tabulated quantities the coefficients of condition. The units used in the tables are the conventional ones adopted in Chapter I. Specifically, n_i and hence $\sigma_{n,i}$ are in units of the earth's radius, as is $\underline{R_i}$, so that $\underline{r_i}$ is dimensionless. $\underline{y}$ is measured in parts in 10^9 per century and $\underline{\dot{n}_M}$ is measured in seconds of arc per century per century ($''/cy^2$).

Aside from the change in parameters, I have made a few other changes between Tables A.I.1 through A.I.17 and the corresponding tables in volume 1. For the records 1133 August 2b B,E and 1133 August 2c B,E in volume 1, the decimal point in σ_n was accidentally shifted.† The positions of the decimal points for these records in Table A.I.12 are the correct ones.

In addition, a number of records imply a magnitude for an eclipse that is less than totality. Hence, they imply a size of n_{obs} that is different from zero, but they do not tell us its sign. For most of these records, we can tell the sign of n_{obs} by calculation without ambiguity. For a few of these records, we cannot tell the sign because the eclipse was too nearly central at the point of observation. These records are those designated as -430 August 3a M, -393 August 14 M, 702 September 26 C, 1153 January 26c E,G, 1290 September 5 E,G, and 1431 February 12c E,I. In volume 1, I failed to notice this problem and tacitly took the sign of n_{obs} to be positive for these records. Here I take these records to be ones that merely imply observation, with no implication about the magnitude. That is, I take the value of n_{obs} for these records to be zero and I take the value of $\sigma_{n,i}$ to be 0.1000.

I use the coefficients of condition in Tables A.I.1 through A.I.17 to infer a value of $\underline{y}$ by the method of least squares, inferring one value of $\underline{y}$ from each table.‡ The values of $\underline{y}$ that I infer, along with their standard deviations, are listed in Table A.I.18 and plotted in Figure A.I.1, with one exception. The values of $\underline{y}$ in Table A.I.18 that apply at the epochs 1446 and 1534 have large standard deviations because the observations are too recent to provide a large "lever arm" for determining an acceleration. As I did in volume 1 (page 445), I replace the values for these two epochs by the single value:

---

†The errors are only in the tabulation. I used the correct values in the analysis.

‡If the reader wishes to test the inference, he should remember that the values of $\underline{r_i}$ in the tables refer to $\underline{y}$ = -20, not to $\underline{y}$ = 0.

$$\underline{y} = -6.0 \pm 16.8 \qquad \text{at the epoch 1486} \qquad \text{(A.I.5)}$$

in plotting Figure A.I.1.

In addition, Figure A.I.1 shows the value $\underline{y} = -9.1 \pm 2.8$ at the epoch 1800. This is obtained from the value $\nu_S' = 1.18 \pm 0.36$ in Table I.4 (page 26) of volume 1, using the relation (Equation I.5) that $\nu_S' = -0.1300\underline{y}$.

I think it is clear from either Table A.I.18 or Figure A.I.1 that $\underline{y}$ has not been constant during the historical period: Every value of $\underline{y}$ for the epoch 1005 or earlier is less than −19 and every value for a later epoch (if we use Equation A.I.5) is greater than −19. This agrees generally with the conclusions of volume 1.

We cannot expect the analysis of the eclipses given here to agree exactly with the analysis in volume 1 for at least three reasons. First, we have changed from the perturbation theories of Newcomb [1895] and Brown [1919] to the DE102 ephemeris. Second, as a consequence of this change, we have had to change the epochs to which the accelerations $\underline{n}_M$ and $\underline{y}$ are referred, and, in fact, we can no longer refer both accelerations to the same epoch. Third, and this is probably the most important reason, we can no longer apply the acceleration perturbations to the mean longitudes of the sun and moon, as we have done in earlier work. Instead, in using the DE102 ephemeris, we of necessity apply the acceleration perturbations to the actual longitudes.

The curve in Figure A.I.1 shows the extent to which the analyses of volumes 1 and 2 agree. This curve gives the value of $\dot{y}$ as a function of time that we derive from the results of volume 1, and which is given in Equation I.19 of this volume. I have not made a formal comparison of the plotted values with the curve, but the curve seems to agree with the points within the uncertainties involved. However, I suspect there is one difference. The curve plotted in the figure has its minimum near the year −1100. If we derive a quadratic from the plotted points in the figure, its minimum would probably be near the year +200. However, the curve is rather flat, and this much change in the position of the minimum is probably not important.

In this volume, I combine the results from the eclipses (Table A.I.18) with the results from other types of observation, and I do this in Section X.1 of the present volume. There is no need to derive a new curve for $\underline{y}$ using only the values in Table A.I.18.

COEFFICIENTS OF CONDITION FOR ECLIPSES
FROM -719 TO -600

Designation			r_i	$n_{Y,i}$	$n_{N,i}$	$\sigma_{n,i}$
- 719 Feb 22	C		-1.197	+ 18.0	- 52.2	0.2830
- 708 Jul 17	C		-1.257	+357.8	-391.7	0.0721
- 694 Oct 10	C		-1.050	+ 21.5	- 71.1	0.2741
- 675 Apr 15	C		-0.594	+ 33.7	+ 22.4	0.2817
- 668 May 27	C		-0.060	- 49.0	+ 38.2	0.1932
- 667 Nov 10	C		+0.364	+ 60.7	- 28.3	0.2754
- 663 Aug 28	C		+0.301	+ 82.8	- 21.8	0.1996
- 654 Aug 19	C		-0.110	+ 87.4	-100.3	0.2850
- 647 Apr 6	C		+1.289	+ 11.6	- 39.1	0.2841
- 625 Feb 3	C		-0.277	- 89.4	+ 17.4	0.2729
- 611 Apr 27	C		+0.503	- 91.4	- 2.2	0.2009
- 601 May 7	C		+1.314	- 70.4	+ 61.9	0.2820
- 600 Sep 20	C		-1.885	+145.7	-294.1	0.0693

TABLE A.I.2

COEFFICIENTS OF CONDITION FOR ECLIPSES
FROM -598 TO -504

Designation	r_i	$\eta_{Y,i}$	$\eta_{N,i}$	$\sigma_{\eta,i}$
- 598 Mar 5 C	-0.772	+ 7.7	+ 34.7	0.2073
- 574 May 9 C	-0.201	+ 29.7	- 35.4	0.2828
- 573 Oct 22 C	+0.574	+ 75.7	- 6.2	0.2828
- 558 Jan 14 C	+1.029	+ 94.8	+105.1	0.2828
- 551 Aug 20 C	+0.415	+ 71.6	- 33.3	0.2828
- 549 Jan 5 C	-0.416	+ 47.9	- 89.6	0.2828
- 548 Jun 19 C	-0.103	+196.0	+ 73.3	0.0707
- 545 Oct 13 C	-0.354	+ 79.1	- 62.5	0.2828
- 534 Mar 18 C	-1.136	- 47.6	+ 13.5	0.2828
- 526 Apr 18 C	+0.539	- 96.5	+122.1	0.2828
- 524 Aug 21 C	+0.134	+ 50.5	- 99.7	0.2000
- 520 Jun 10 C	+1.108	- 6.4	+ 2.9	0.2000
- 519 Nov 23 C	-0.825	+ 5.7	+ 2.5	0.2828
- 517 Apr 9 C	+0.879	- 72.9	+ 7.6	0.2828
- 510 Nov 14 C	+0.527	+102.9	- 95.5	0.2828
- 504 Feb 16 C	+1.445	- 74.8	+104.3	0.2828

TABLE A.I.3

COEFFICIENTS OF CONDITION FOR ECLIPSES
FROM -497 TO -299

Designation	r_i	$\eta_{Y,i}$	$\eta_{N,i}$	$\sigma_{\eta,i}$
- 497 Sep 22 C	+0.017	+ 81.8	- 33.7	0.2828
- 494 Jul 22 C	-0.932	+ 50.1	+ 17.1	0.2828
- 480 Apr 19 C	-0.277	- 34.9	+ 10.0	0.2828
- 430 Aug 3a M	+0.068	+109.8	-193.2	0.1000
- 430 Aug 3b M	+0.060	+ 37.1	- 64.8	0.3007
- 423 Mar 21 M	+0.604	-202.0	- 8.6	0.0790
- 393 Aug 14 M	-0.076	+204.0	-166.8	0.1000
- 381 Jul 3 C	+0.964	-102.3	+102.3	0.1971
- 363 Jul 13 M	+0.650	- 23.7	+ 46.7	0.3431
- 309 Aug 15a M	+0.255	- 29.3	+ 87.6	0.1347
- 299 Jul 26 C	+0.298	- 47.8	+ 6.5	0.1917

TABLE A.I.4

COEFFICIENTS OF CONDITION FOR ECLIPSES
FROM -197 TO -135

Designation	r_i	$\eta_{Y,i}$	$\eta_{N,i}$	$\sigma_{\eta,i}$
- 197 Aug 7 C	+0.392	+ 35.7	+188.9	0.0561
- 187 Jul 17 C	-2.014	+280.7	-316.8	0.0642
- 180 Mar 4 C	+0.615	-181.5	+ 46.3	0.0626
- 146 Nov 10 C	-0.112	+101.1	- 30.9	0.0880
- 135 Apr 15 BA	+1.174	-651.4	+543.4	0.0311

TABLE A.I.5

COEFFICIENTS OF CONDITION FOR ECLIPSES
FROM -88 TO -1

Designation			r_i	$\eta_{Y,i}$	$\eta_{N,i}$	$\sigma_{\eta,i}$
- 88 Sep	29	C	+0.607	+ 50.4	+ 25.7	0.0793
- 79 Sep	20	C	+0.306	+277.5	-303.8	0.0661
- 27 Jun	19	C	-0.155	-276.0	+132.7	0.0645
- 15 Nov	1	C	+1.199	- 26.0	+ 39.5	0.2000
- 14 Mar	29	C	-1.554	- 96.3	+ 10.3	0.1184
- 1 Feb	5	C	+0.408	- 38.5	+ 59.1	0.0775

TABLE A.I.6

COEFFICIENTS OF CONDITION FOR ECLIPSES
FROM 2 TO 243

Designation			r_i	$\eta_{Y,i}$	$\eta_{N,i}$	$\sigma_{\eta,i}$
2 Nov	23	C	+0.051	+188.2	- 60.3	0.0627
59 Apr	30	E	-0.931	+ 19.2	- 31.9	0.1001
59 Apr	30	M	-1.383	+ 60.7	- 82.1	0.1007
65 Dec	16	C	-0.093	+155.8	-220.7	0.0627
118 Sep	3	E,I	+0.394	+ 28.8	+ 10.3	0.2370
120 Jan	18	C	+0.252	-187.1	+ 19.1	0.0587
243 Jun	5	C	+1.175	- 50.3	+ 17.8	0.1934

COEFFICIENTS OF CONDITION FOR ECLIPSES
FROM 346 TO 484

Designation	$r_{\underline{i}}$	$\eta_{\underline{Y},\underline{i}}$	$\eta_{\underline{N},\underline{i}}$	$\sigma_{\eta,\underline{i}}$
346 Jun 6a M,B	−0.649	− 82.1	+ 85.8	0.0977
348 Oct 9 M,B	+0.651	+ 36.6	− 33.9	0.2236
360 Aug 28 C	−0.239	+ 41.2	− 1.6	0.1907
393 Nov 20 E,I	+0.073	+ 16.4	− 1.9	0.2256
402 Nov 11 E,SP	+0.209	+ 57.5	− 56.5	0.1444
418 Jul 19a E,I	−0.127	+ 19.7	+ 18.5	0.1518
418 Jul 19 E,SP	−0.047	− 5.7	+ 24.8	0.2272
418 Jul 19a M,B	−0.664	+137.1	+ 1.4	0.0437
418 Jul 19c M,B	−0.205	+ 42.4	+ 0.4	0.1414
429 Dec 12 C	+0.410	− 38.0	+ 41.0	0.0722
447 Dec 23 E,SP	+0.079	− 61.0	+ 62.3	0.1007
458 May 28 E,SP	+0.287	−101.7	+136.6	0.0666
464 Jul 20 E,SP	−0.002	− 57.7	+ 6.3	0.0555
484 Jan 14 M,B	+0.070	+ 55.0	− 7.1	0.0618

TABLE A.I.8

COEFFICIENTS OF CONDITION FOR ECLIPSES
FROM 512 TO 693

Designation			r_i	$\eta_{Y,i}$	$\eta_{N,i}$	$\sigma_{n,i}$
512 Jun 29	E,I		−0.639	− 34.4	+ 50.4	0.1414
516 Apr 18	C		−0.223	− 31.8	+ 39.4	0.1734
522 Jun 10	C		−0.374	− 30.0	+ 7.8	0.1971
534 Apr 29	E,I		−0.643	− 23.7	+ 30.8	0.2286
538 Feb 15	E,I		−0.258	− 3.1	+ 9.7	0.4536
540 Jun 20	E,I		−0.124	− 10.8	+ 0.3	0.2796
563 Oct 3	E,F		+1.305	+ 30.0	+ 14.0	0.0899
590 Oct 4	E,F		+0.800	+119.2	−149.4	0.0636
590 Oct 4	M,B		−0.416	+119.2	−191.0	0.0490
592 Mar 19	E,F		+0.243	− 32.0	+ 32.9	0.1886
601 Mar 10	Is,E		+0.585	− 59.8	+ 10.2	0.1000
603 Aug 12	E,F		−0.435	+ 26.1	− 2.9	0.1426
644 Nov 5	M,B		−0.150	+ 26.2	− 53.0	0.1414
655 Apr 12	E,SP		+0.080	− 33.6	− 9.6	0.0783
664 May 1	B,E		+0.176	+ 14.2	+ 7.5	0.1465
664 May 1a	B,I		+0.195	+ 11.5	+ 10.1	0.1458
664 May 1b	B,I		+0.062	+ 3.6	+ 3.2	0.4611
688 Jul 3	B,I		+0.387	− 6.9	− 4.0	0.1838
693 Oct 5a	M,B		−0.299	+ 63.9	− 5.8	0.0618
693 Oct 5b	M,B		−0.299	+ 63.9	− 5.8	0.0618

COEFFICIENTS OF CONDITION FOR ECLIPSES
FROM 702 TO 841

Designation		$\underline{r}_{\underline{i}}$	$\eta_{\underline{Y},\underline{i}}$	$\eta_{\underline{N},\underline{i}}$	$\sigma_{\eta,\underline{i}}$
702 Sep 26	C	+0.020	+ 25.9	− 47.3	0.1000
729 Oct 27	C	−0.797	+ 45.8	− 6.7	0.0860
733 Aug 14a	B,E	−0.050	+ 0.9	+ 17.6	0.1453
733 Aug 14b	B,E	+0.333	+ 2.0	+ 32.1	0.0784
753 Jan 9	B,E	−0.471	− 2.3	− 17.5	0.1475
753 Jan 9	B,I	−0.269	+ 1.1	− 14.0	0.2245
754 Jun 25	C	+0.287	+ 1.1	+ 35.4	0.0820
760 Aug 15	E,CE	+0.194	+ 9.1	− 15.4	0.3162
760 Aug 15	M,B	+0.306	+ 14.1	− 29.3	0.1414
761 Aug 5	C	−0.179	+ 63.6	− 68.0	0.0610
764 Jun 4	B,I	−0.045	− 14.6	− 4.1	0.1457
764 Jun 4	E,F	+0.721	− 0.4	− 9.6	0.1414
764 Jun 4	E,G	+0.495	+ 3.0	− 13.2	0.1414
764 Jun 4	E,CE	+0.270	+ 4.0	− 8.4	0.3162
787 Sep 16a	E,G	−1.471	+ 11.5	+ 13.5	0.1000
787 Sep 16b	E,G	−1.291	+ 8.8	+ 8.8	0.1414
787 Sep 16	E,CE	−0.436	+ 5.0	+ 3.2	0.3162
787 Sep 16	M,B	−1.404	+ 57.8	0.0	0.0490
807 Feb 11	B,W	+0.162	− 6.4	− 5.1	0.2236
807 Feb 11	E,F	+0.430	− 14.4	− 3.9	0.1414
807 Feb 11	E,G	+0.886	− 25.0	− 0.3	0.1000
807 Feb 11	E,I	+1.171	− 23.3	+ 4.9	0.1414
807 Feb 11a	E,CE	+0.849	− 20.4	+ 2.8	0.1414
807 Feb 11b	E,CE	+0.400	− 9.4	+ 1.6	0.3162
808 Jul 27	C	−0.558	− 9.2	+ 18.5	0.2000
809 Jul 16	B,E	+0.981	− 6.5	+ 22.2	0.1459
810 Nov 30	E,G	+0.343	+ 25.1	− 40.8	0.1000
810 Nov 30a	E,CE	+0.185	+ 15.8	− 27.9	0.1414
810 Nov 30b	E,CE	+0.042	+ 6.2	− 11.8	0.3162
812 May 14a	E,G	−1.448	− 9.5	+ 23.7	0.1414

Designation	r_i	$\eta_{Y,i}$	$\eta_{N,i}$	$\sigma_{\eta,i}$
812 May 14b E,G	−1.528	− 8.5	+ 22.8	0.1414
812 May 14 M,B	−0.624	+ 12.2	+ 24.5	0.0681
813 May 4 M,B	+0.373	− 18.8	+ 29.2	0.1000
818 Jul 7a E,G	+1.758	− 18.3	+ 4.0	0.1414
818 Jul 7b E,G	+1.715	− 17.8	+ 3.5	0.1414
840 May 5a E,F	−0.262	− 5.1	− 5.9	0.1414
840 May 5b E,F	+0.062	− 13.5	− 18.9	0.0465
840 May 5c E,F	−0.436	− 3.4	− 11.3	0.1000
840 May 5d E,F	−0.042	− 10.5	− 5.8	0.1000
840 May 5a E,G	−0.099	− 1.2	− 5.3	0.2236
840 May 5b E,G	−0.099	− 1.2	− 5.4	0.2236
840 May 5c E,G	−1.492	− 4.8	− 28.6	0.0437
840 May 5d E,G	−1.943	− 10.1	− 24.9	0.0437
840 May 5e E,G	−0.817	− 1.8	− 29.7	0.0454
840 May 5a E,I	+0.109	+ 0.1	− 9.8	0.1414
840 May 5b E,I	+0.639	+ 0.2	− 28.0	0.0490
840 May 5c E,I	+0.209	− 4.6	− 26.7	0.0454
841 Oct 18 E,F	−2.691	+ 21.1	+ 2.1	0.1002

COEFFICIENTS OF CONDITION FOR ECLIPSES
FROM 865 TO 891

Designation	r_i	$\eta_{Y,i}$	$\eta_{N,i}$	$\sigma_{\eta,i}$
865 Jan 1 B,I	+0.012	− 10.0	− 5.4	0.1422
878 Oct 29 B,E	+0.032	+ 6.3	+ 8.7	0.1006
878 Oct 29 B,I	+0.074	+ 6.8	+ 4.4	0.1423
878 Oct 29 B,W	+0.064	+ 3.9	+ 3.0	0.2236
878 Oct 29a E,F	+0.149	+ 2.6	+ 3.9	0.2241
878 Oct 29b E,F	+0.119	+ 3.7	+ 10.6	0.1000
878 Oct 29c E,F	+0.765	+ 5.0	+ 33.5	0.0358
878 Oct 29d E,F	+0.278	+ 3.9	+ 10.3	0.1000
878 Oct 29e E,F	+0.204	+ 6.5	+ 21.7	0.0506
878 Oct 29f E,F	+0.069	+ 2.8	+ 7.4	0.1414
878 Oct 29a E,G	+0.171	− 0.5	+ 9.9	0.1414
878 Oct 29b E,G	+0.171	− 0.5	+ 9.9	0.1414
878 Oct 29c E,G	+0.193	− 1.9	+ 43.9	0.0321
878 Oct 29d E,G	+0.095	+ 1.3	+ 12.5	0.1000
885 Jun 16 B,I	+0.444	− 23.9	+ 35.1	0.0960
885 Jun 16 B,SM	+0.175	− 13.5	+ 21.9	0.1414
888 Apr 15 C	−1.751	− 78.1	+103.3	0.0699
891 Aug 8 E,F	−0.382	+ 9.1	− 18.5	0.1414
891 Aug 8 E,CE	−0.143	+ 14.8	− 23.7	0.1414
891 Aug 8a E,CE	−0.282	+ 17.0	− 26.0	0.1414
891 Aug 8a M,B	−0.756	+ 85.1	−112.5	0.0505
891 Aug 8b M,B	−0.756	+ 85.1	−112.5	0.0505
891 Aug 8c M,B	−0.391	+ 44.0	− 58.1	0.0977

TABLE A.I.11

COEFFICIENTS OF CONDITION FOR ECLIPSES
FROM 912 TO 1098

Designation		r_i	$\eta_{Y,i}$	$\eta_{N,i}$	$\sigma_{\eta,i}$
912 Jun 17	E,SP	+0.092	+ 34.1	− 52.3	0.0618
939 Jul 19	E,G	−0.223	− 14.9	+ 31.0	0.0783
939 Jul 19a	E,I	−0.839	− 17.8	+ 33.7	0.0783
939 Jul 19c	E,I	−0.039	− 10.3	+ 19.0	0.1414
939 Jul 19	E,CE	−0.618	− 13.5	+ 25.8	0.1000
961 May 17a	E,G	−0.019	− 26.4	+ 39.2	0.1000
961 May 17b	E,G	−0.019	− 26.4	+ 39.2	0.1000
968 Dec 22	E,F	−0.230	+ 24.8	+ 6.5	0.0448
968 Dec 22	E,G	−0.182	+ 5.1	+ 4.0	0.1414
968 Dec 22a	E,I	+0.059	+ 3.4	+ 4.9	0.1414
968 Dec 22b	E,I	+0.083	+ 3.5	+ 4.8	0.1414
968 Dec 22	E,CE	−0.246	+ 7.1	+ 5.7	0.1000
968 Dec 22a	E,BN	−0.243	+ 5.8	+ 3.6	0.1436
968 Dec 22b	E,BN	−0.078	+ 1.9	+ 1.2	0.4472
968 Dec 22a	M,B	+0.243	− 26.2	+ 53.0	0.0321
968 Dec 22b	M,B	+0.126	− 13.6	+ 27.5	0.0618
968 Dec 22c	M,B	+0.078	− 8.4	+ 17.0	0.1000
990 Oct 21a	E,G	+0.578	+ 18.2	− 5.0	0.1000
990 Oct 21b	E,G	+0.601	+ 18.2	− 4.9	0.1000
990 Oct 21c	E,G	+0.388	+ 12.8	− 3.6	0.1418
1009 Mar 29a	E,BN	+2.858	− 9.8	− 4.9	0.1000
1009 Mar 29b	E,BN	+1.974	− 6.6	− 3.9	0.1414
1018 Apr 18	E,G	−0.338	+ 11.1	+ 2.3	0.1111
1018 Apr 18	E,BN	−0.897	+ 3.4	+ 1.2	0.3162
1023 Jan 24	B,E	+0.171	− 20.6	+ 27.6	0.1000
1023 Jan 24	B,I	−0.101	− 12.8	+ 17.9	0.1426
1023 Jan 24	E,F	+0.543	− 18.2	+ 22.3	0.1414
1023 Jan 24	E,BN	+0.401	− 16.3	+ 21.1	0.1414
1030 Aug 31	B,I	+0.894	+ 11.2	− 1.8	0.1441
1033 Jun 29a	E,F	+0.940	− 1.8	+ 24.4	0.0665

Designation	r_i	$\eta_{Y,i}$	$\eta_{N,i}$	$\sigma_{n,i}$
1033 Jun 29b E,F	-0.420	+ 0.8	+ 14.2	0.1000
1033 Jun 29c E,F	-0.166	- 1.1	+ 16.6	0.0977
1033 Jun 29d E,F	-1.513	- 4.3	+ 38.2	0.0437
1033 Jun 29e E,F	-0.124	- 2.3	+ 17.5	0.1000
1033 Jun 29f E,F	-0.175	- 1.4	+ 12.1	0.1414
1033 Jun 29a E,G	-0.903	+ 1.9	+ 13.0	0.1000
1033 Jun 29b E,G	-0.210	+ 1.7	+ 13.6	0.1000
1033 Jun 29c E,G	-0.381	+ 1.6	+ 13.5	0.1000
1033 Jun 29a E,I	+1.169	+ 10.0	+ 13.3	0.0693
1033 Jun 29b E,I	-0.253	+ 5.5	+ 9.0	0.1107
1033 Jun 29c E,I	+0.103	+ 1.3	+ 9.6	0.1414
1033 Jun 29 E,BN	-0.529	- 1.2	+ 11.7	0.1414
1037 Apr 18a E,F	+1.390	- 40.3	+ 58.4	0.0618
1037 Apr 18b E,F	+0.942	- 41.1	+ 59.1	0.0618
1037 Apr 18 E,BN	+0.763	- 16.7	+ 24.8	0.1414
1039 Aug 22a E,F	-0.740	+ 26.9	- 36.9	0.1000
1039 Aug 22b E,F	-0.369	+ 28.5	- 38.3	0.1000
1039 Aug 22a E,G	-0.427	+ 12.3	- 16.7	0.2236
1039 Aug 22b E,G	-1.161	+ 26.1	- 36.2	0.1000
1039 Aug 22c E,G	-1.236	+ 27.4	- 37.8	0.1000
1039 Aug 22 E,BN	-0.567	+ 17.7	- 24.7	0.1414
1044 Nov 22a E,F	-0.389	+ 12.8	- 2.1	0.1000
1044 Nov 22b E,F	-0.296	+ 9.1	- 1.4	0.1414
1044 Nov 22c E,F	-0.400	+ 12.3	- 1.7	0.1000
1044 Nov 22a E,G	-0.835	+ 8.4	+ 2.2	0.1000
1044 Nov 22b E,G	-0.802	+ 7.5	+ 3.0	0.1000
1044 Nov 22c E,G	-0.807	+ 7.1	+ 3.4	0.1000
1044 Nov 22 E,I	-0.301	+ 9.1	+ 0.8	0.1000
1079 Jul 1 E,SP	+0.230	+ 62.1	- 81.5	0.0330
1087 Aug 1 E,CE	-2.285	- 3.6	+ 12.0	0.1414

Designation	r_i	$\eta_{Y,i}$	$\eta_{N,i}$	$\sigma_{\eta,i}$
1093 Sep 23a B,E	+0.312	+ 6.4	− 9.7	0.2957
1093 Sep 23b B,E	+0.190	+ 6.0	− 9.4	0.2957
1093 Sep 23a E,G	+0.164	+ 21.5	− 31.7	0.1000
1093 Sep 23b E,G	+0.501	+ 24.1	− 34.1	0.1000
1093 Sep 23c E,G	+0.439	+ 16.9	− 24.0	0.1414
1093 Sep 23d E,G	+0.621	+ 23.9	− 34.0	0.1000
1093 Sep 23e E,G	+0.336	+ 23.0	− 33.2	0.1000
1093 Sep 23 E,I	−0.050	+ 26.0	− 35.0	0.1107
1093 Sep 23 E,BN	+0.701	+ 21.0	− 31.0	0.1000
1093 Sep 23a E,CE	+0.138	+ 11.1	− 15.7	0.2236
1093 Sep 23b E,CE	+0.064	+ 23.3	− 33.6	0.1000
1093 Sep 23c E,CE	+1.696	+ 55.1	− 78.0	0.0437
1098 Dec 25 E,F	−1.308	− 5.4	+ 15.3	0.1000
1098 Dec 25a E,G	−1.035	− 11.4	+ 21.8	0.1000

COEFFICIENTS OF CONDITION FOR ECLIPSES
FROM 1109 TO 1147

Designation	r_i	$\eta_{Y,i}$	$\eta_{N,i}$	$\sigma_{n,i}$
1109 May 31 E,BN	+1.359	− 2.0	+ 14.8	0.1000
1113 Mar 19 M,HL	+0.308	− 16.0	+ 27.0	0.0636
1118 May 22 E,BN	+2.696	− 13.5	+ 3.9	0.1000
1124 Aug 11a B,E	+0.329	+ 4.2	+ 0.4	0.2476
1124 Aug 11b B,E	+0.319	+ 4.1	+ 0.5	0.2476
1124 Aug 11c B,E	+0.323	+ 4.0	+ 0.6	0.2476
1124 Aug 11a E,G	+1.502	+ 11.9	− 0.6	0.1000
1124 Aug 11b E,G	+1.267	+ 11.4	− 0.1	0.1000
1124 Aug 11 E,BN	+0.947	+ 14.7	0.0	0.0791
1124 Aug 11 E,CE	+1.452	+ 13.3	− 1.9	0.1000
1124 Aug 11 M,HL	+1.690	+ 22.2	− 8.0	0.0783
1133 Aug 2a B,E	−0.221	+ 17.2	− 24.6	0.0874
1133 Aug 2b B,E	+0.830	+ 22.9	− 33.4	0.0629
1133 Aug 2c B,E	+0.511	+ 22.3	− 33.2	0.0624
1133 Aug 2d B,E	+0.550	+ 23.8	− 34.5	0.0629
1133 Aug 2 B,I	+0.342	+ 8.1	− 12.8	0.1466
1133 Aug 2 B,SM	+0.019	+ 12.9	− 20.1	0.0963
1133 Aug 2a E,F	+0.879	+ 36.2	− 51.5	0.0437
1133 Aug 2b E,F	+0.480	+ 16.9	− 23.4	0.1000
1133 Aug 2c E,F	+0.975	+ 37.8	− 52.6	0.0437
1133 Aug 2d E,F	+2.249	+ 53.8	− 71.9	0.0342
1133 Aug 2e E,F	+0.643	+ 16.3	− 22.8	0.1000
1133 Aug 2f E,F	+0.288	+ 7.3	− 10.2	0.2236
1133 Aug 2b E,G	−0.183	+ 38.7	− 53.5	0.0437
1133 Aug 2c E,G	+0.063	+ 13.2	− 17.8	0.1414
1133 Aug 2d E,G	+0.199	+ 13.3	− 17.8	0.1414
1133 Aug 2e E,G	+0.551	+ 52.6	− 72.6	0.0321
1133 Aug 2f E,G	−0.003	+ 12.8	− 17.5	0.1414
1133 Aug 2g E,G	−0.309	+ 38.7	− 54.0	0.0437
1133 Aug 2h E,G	+0.327	+ 28.6	− 39.2	0.0618

Designation	$r_{\underline{i}}$	$n_{\underline{Y},\underline{i}}$	$n_{\underline{N},\underline{i}}$	$\sigma_{n,\underline{i}}$
1133 Aug 2i E,G	+0.045	+ 13.0	- 17.6	0.1414
1133 Aug 2j E,G	+0.196	+ 57.3	- 77.6	0.0321
1133 Aug 2k E,G	+0.209	+ 56.0	- 76.6	0.0321
1133 Aug 2ℓ E,G	-0.163	+ 12.4	- 17.2	0.1414
1133 Aug 2m E,G	+0.345	+ 28.5	- 39.2	0.0618
1133 Aug 2n E,G	+0.017	+ 18.8	- 25.2	0.1000
1133 Aug 2 E,I	-0.395	+ 29.1	- 36.8	0.0783
1133 Aug 2a E,BN	+0.668	+ 38.2	- 53.1	0.0437
1133 Aug 2b E,BN	+0.716	+ 38.2	- 53.3	0.0437
1133 Aug 2c E,BN	+0.291	+ 16.2	- 22.8	0.1000
1133 Aug 2d E,BN	+0.535	+ 38.7	- 53.3	0.0437
1133 Aug 2e E,BN	+0.137	+ 48.6	- 69.2	0.0321
1133 Aug 2a E,CE	-0.206	+ 35.2	- 46.9	0.0554
1133 Aug 2b E,CE	+0.280	+ 60.7	- 80.7	0.0321
1133 Aug 2c E,CE	+0.050	+ 61.4	- 81.6	0.0321
1133 Aug 2f E,CE	+0.008	+ 20.0	- 26.7	0.0963
1133 Aug 2g E,CE	-0.183	+ 13.2	- 17.8	0.1414
1133 Aug 2h E,CE	+0.677	+ 42.1	- 57.4	0.0437
1133 Aug 2i E,CE	-0.348	+ 19.2	- 25.8	0.1000
1140 Mar 20a B,E	+0.147	- 7.4	- 2.8	0.0963
1140 Mar 20b B,E	+0.198	- 12.6	- 3.3	0.0602
1140 Mar 20c B,E	+0.415	- 11.6	- 4.5	0.0602
1140 Mar 20d B,E	+0.036	- 7.6	- 2.2	0.1000
1140 Mar 20e B,E	+0.150	- 7.5	- 2.5	0.0963
1140 Mar 20f B,E	+0.158	- 6.9	- 2.6	0.1000
1140 Mar 20g B,E	+0.025	- 5.4	- 1.6	0.1414
1140 Mar 20 B,SM	-0.306	- 5.0	- 1.9	0.1414
1140 Mar 20 B,W	-0.011	- 5.9	- 1.1	0.1414
1140 Mar 20 E,F	+0.231	- 4.7	- 2.0	0.1414
1140 Mar 20 E,G	+0.928	- 5.6	- 3.7	0.1000

Designation	r_i	$n_{Y,i}$	$n_{N,i}$	$\sigma_{n,i}$
1140 Mar 20a E,BN	+1.374	− 19.0	− 10.0	0.0321
1140 Mar 20b E,BN	+0.716	− 13.1	− 6.5	0.0490
1140 Mar 20 E,PR	+0.235	− 3.6	− 5.6	0.1000
1140 Mar 20a E,Sc	−0.119	− 3.8	− 5.4	0.1000
1140 Mar 20b E,Sc	−0.234	− 6.0	− 7.0	0.0717
1140 Mar 20c E,Sc	−0.234	− 6.0	− 7.0	0.0717
1147 Oct 26a E,F	+0.541	+ 18.9	− 29.2	0.1000
1147 Oct 26b E,F	+0.420	+ 18.9	− 29.3	0.1000
1147 Oct 26c E,F	+0.718	+ 20.2	− 30.5	0.1000
1147 Oct 26b E,G	+0.636	+ 40.5	− 64.8	0.0437
1147 Oct 26c E,G	+0.133	+ 18.4	− 28.9	0.1000
1147 Oct 26d E,G	+0.120	+ 19.0	− 29.5	0.1000
1147 Oct 26e E,G	+0.061	+ 18.4	− 28.8	0.1000
1147 Oct 26f E,G	+0.192	+ 14.1	− 21.6	0.1414
1147 Oct 26a E,BN	+0.568	+ 42.6	− 65.9	0.0437
1147 Oct 26b E,BN	+0.416	+ 42.3	− 66.1	0.0437
1147 Oct 26c E,BN	+0.059	+ 17.5	− 27.8	0.1000
1147 Oct 26 E,CE	+0.241	+ 14.5	− 21.9	0.1414
1147 Oct 26 M,B	−1.082	+ 24.0	− 39.1	0.0783

TABLE A.I.13

COEFFICIENTS OF CONDITION FOR ECLIPSES
FROM 1153 TO 1194

Designation	r_i	$n_{Y,i}$	$n_{N,i}$	$\sigma_{n,i}$
1153 Jan 26a E,F	-0.580	- 16.5	+ 27.3	0.1000
1153 Jan 26b E,F	-0.756	- 16.1	+ 26.6	0.1000
1153 Jan 26c E,F	-0.559	- 16.7	+ 27.4	0.1000
1153 Jan 26a E,G	-0.023	- 18.6	+ 29.8	0.1000
1153 Jan 26b E,G	+0.124	- 18.7	+ 30.1	0.1000
1153 Jan 26c E,G	-0.117	- 17.4	+ 28.7	0.1000
1153 Jan 26d E,G	-0.375	- 16.9	+ 27.9	0.1000
1153 Jan 26 E,CE	+0.271	- 13.7	+ 21.8	0.1414
1163 Jul 3 E,I	-1.485	- 13.3	+ 24.7	0.1000
1178 Sep 13a B,E	-0.762	+ 10.9	- 1.3	0.1000
1178 Sep 13b B,E	+0.328	+ 14.2	- 1.9	0.0783
1178 Sep 13c B,E	-0.526	+ 7.5	- 0.7	0.1414
1178 Sep 13 B,SM	-1.066	+ 21.5	0.0	0.0437
1178 Sep 13a B,W	-0.305	+ 4.6	- 0.3	0.2242
1178 Sep 13b B,W	-0.273	+ 4.7	- 0.4	0.2236
1178 Sep 13a E,F	-0.195	+ 8.7	- 1.8	0.1414
1178 Sep 13b E,F	-0.780	+ 11.5	- 1.9	0.1000
1178 Sep 13c E,F	+0.808	+ 27.9	- 5.7	0.0437
1178 Sep 13d E,F	-0.540	+ 12.6	- 2.9	0.1000
1178 Sep 13e E,F	+1.144	+ 29.7	- 7.6	0.0437
1178 Sep 13f E,F	+0.604	+ 26.8	- 4.8	0.0437
1178 Sep 13g E,F	+0.118	+ 19.6	- 5.5	0.0693
1178 Sep 13h E,F	-0.122	+ 9.1	- 2.2	0.1414
1178 Sep 13i E,F	+0.580	+ 28.0	- 6.7	0.0460
1178 Sep 13 E,G	-0.649	+ 8.6	- 1.9	0.1414
1178 Sep 13a E,I	-0.347	+ 14.2	- 4.7	0.1000
1178 Sep 13b E,I	+0.893	+ 18.1	- 6.0	0.0783
1178 Sep 13c E,I	-0.487	+ 9.3	-. 2.5	0.1414
1178 Sep 13d E,I	-0.230	+ 10.0	- 3.3	0.1414
1178 Sep 13e E,I	+0.465	+ 31.4	- 9.4	0.0437

Designation	r_i	$\eta_{Y,i}$	$\eta_{N,i}$	$\sigma_{n,i}$
1178 Sep 13 E,Sc	−1.144	+ 6.8	− 0.4	0.1426
1180 Jan 28 E,F	+0.708	− 10.4	+ 2.4	0.1000
1181 Jul 13 E,F	+0.028	+ 6.5	− 2.2	0.1800
1185 May 1a B,E	+1.080	− 5.2	+ 13.5	0.1000
1185 May 1b B,E	+0.126	− 6.1	+ 16.7	0.0783
1185 May 1c B,E	+1.043	− 4.9	+ 13.2	0.1000
1185 May 1d B,E	+0.496	− 2.5	+ 6.2	0.2236
1185 May 1e B,E	+0.912	− 5.1	+ 13.4	0.1000
1185 May 1a B,SM	+0.862	− 11.8	+ 29.9	0.0458
1185 May 1b B,SM	+0.393	− 4.5	+ 10.0	0.1414
1185 May 1a B,W	−0.093	− 3.9	+ 8.6	0.1751
1185 May 1b B,W	−0.038	− 3.7	+ 8.4	0.1751
1185 May 1a E,F	+1.240	− 4.1	+ 10.0	0.1414
1185 May 1c E,F	+0.639	− 5.9	+ 16.7	0.0783
1185 May 1 E,Sc	+0.397	− 1.0	+ 5.8	0.1769
1186 Apr 21a B,E	+3.114	− 2.4	+ 10.1	0.1000
1186 Apr 21b B,E	+2.665	− 3.4	+ 13.0	0.0783
1187 Sep 4 B,E	+0.650	+ 19.3	− 27.1	0.0783
1187 Sep 4 E,F	+0.898	+ 20.9	− 28.6	0.0783
1187 Sep 4a E,G	+1.205	+ 17.6	− 23.7	0.1000
1187 Sep 4b E,G	+1.177	+ 17.8	− 23.8	0.1000
1187 Sep 4c E,G	+0.902	+ 16.1	− 22.3	0.1000
1187 Sep 4d E,G	+0.829	+ 16.4	− 22.6	0.1000
1187 Sep 4e E,G	+1.213	+ 16.1	− 22.2	0.1000
1187 Sep 4f E,G	+1.204	+ 17.9	− 23.9	0.1000
1187 Sep 4g E,G	−0.599	+ 22.0	− 30.0	0.0749
1187 Sep 4a E,I	+1.136	+ 13.5	− 17.6	0.1414
1187 Sep 4b E,I	+1.156	+ 13.4	− 17.5	0.1414
1187 Sep 4 E,BN	+1.416	+ 15.6	− 21.7	0.1000
1187 Sep 4a E,CE	+0.886	+ 17.9	− 24.0	0.1000

Designation	r_i	$\eta_{Y,i}$	$\eta_{N,i}$	$\sigma_{\eta,i}$
1187 Sep 4b E,CE	+1.100	+ 18.0	− 24.0	0.1000
1187 Sep 4c E,CE	+1.605	+ 29.1	− 38.8	0.0618
1187 Sep 4a E,Sc	+0.349	+ 9.8	− 14.4	0.1433
1187 Sep 4 M,HL	+0.579	+ 19.9	− 26.5	0.1000
1191 Jun 23a B,E	−0.664	+ 2.3	− 10.0	0.1107
1191 Jun 23b B,E	−0.687	+ 2.7	− 10.4	0.1107
1191 Jun 23c B,E	+0.179	+ 1.1	− 4.8	0.2236
1191 Jun 23d B,E	+0.361	+ 4.8	− 18.4	0.0626
1191 Jun 23e B,E	−0.789	+ 2.8	− 13.5	0.0783
1191 Jun 23f B,E	+0.415	+ 5.6	− 25.1	0.0443
1191 Jun 23 B,SM	−0.259	+ 1.5	− 9.9	0.1000
1191 Jun 23 B,W	+0.162	+ 0.5	− 4.2	0.2236
1191 Jun 23a E,F	+0.627	+ 9.4	− 29.3	0.0437
1191 Jun 23b E,F	+0.990	+ 2.8	− 11.1	0.1000
1191 Jun 23c E,F	+0.921	+ 3.6	− 12.0	0.1000
1191 Jun 23d E,F	+0.630	+ 5.6	− 17.3	0.0745
1191 Jun 23e E,F	+0.624	+ 5.6	− 14.5	0.1000
1191 Jun 23f E,F	+0.724	+ 4.0	− 12.5	0.1000
1191 Jun 23g E,F	+0.311	+ 2.6	− 8.7	0.1414
1191 Jun 23a E,G	+0.481	+ 5.1	− 11.5	0.1414
1191 Jun 23b E,G	+0.925	+ 11.7	− 25.6	0.0681
1191 Jun 23c E,G	+0.413	+ 5.7	− 12.3	0.1414
1191 Jun 23d E,G	+0.329	+ 5.5	− 14.5	0.1000
1191 Jun 23e E,G	−0.241	+ 6.0	− 14.1	0.1107
1191 Jun 23f E,G	+0.799	+ 10.4	− 25.1	0.0618
1191 Jun 23g E,G	+0.222	+ 7.1	− 16.3	0.1000
1191 Jun 23a E,I	+0.717	+ 6.5	− 13.1	0.1414
1191 Jun 23b E,I	+0.844	+ 6.4	− 12.9	0.1414
1191 Jun 23a E,BN	+0.455	+ 4.7	− 13.6	0.1000
1191 Jun 23b E,BN	+0.403	+ 5.1	− 13.9	0.1000

Designation	$r_{\underline{i}}$	$\eta_{\underline{Y},\underline{i}}$	$\eta_{\underline{N},\underline{i}}$	$\sigma_{\eta,\underline{i}}$
1191 Jun 23c E,BN	+0.283	+ 3.1	− 9.3	0.1414
1191 Jun 23a E,CE	+0.483	+ 9.4	− 19.0	0.1000
1191 Jun 23b E,CE	+0.630	+ 8.6	− 18.2	0.1000
1191 Jun 23c E,CE	+0.389	+ 6.4	− 13.2	0.1414
1191 Jun 23d E,CE	+0.539	+ 8.6	− 18.2	0.1000
1194 Apr 22 B,E	+1.839	− 1.3	− 8.6	0.0783

COEFFICIENTS OF CONDITION FOR ECLIPSES
FROM 1207 TO 1290

Designation	r_i	$^n Y,i$	$^n N,i$	$\sigma_{n,i}$
1207 Feb 28a B,E	-0.630	- 10.6	+ 17.7	0.1414
1207 Feb 28b B,E	-0.600	- 10.6	+ 17.6	0.1414
1207 Feb 28c B,E	-0.608	- 10.7	+ 17.8	0.1414
1207 Feb 28d B,E	-0.600	- 10.6	+ 17.6	0.1414
1207 Feb 28 E,F	-0.460	- 11.8	+ 18.9	0.1414
1207 Feb 28a E,F	+0.658	- 21.1	+ 34.0	0.0783
1207 Feb 28b E,F	+1.020	- 23.6	+ 36.5	0.0783
1207 Feb 28c E,F	-0.260	- 16.8	+ 27.0	0.1000
1207 Feb 28a E,G	+0.057	- 17.3	+ 27.8	0.1000
1207 Feb 28b E,G	+0.221	- 17.3	+ 27.8	0.1000
1207 Feb 28c E,G	-0.333	- 16.0	+ 26.3	0.1000
1207 Feb 28d E,G	+0.251	- 17.6	+ 28.0	0.1000
1207 Feb 28e E,G	+0.137	- 17.4	+ 27.8	0.1000
1207 Feb 28 E,PR	+0.223	- 12.1	+ 22.1	0.1012
1207 Feb 28 E,BN	-0.570	- 15.8	+ 25.9	0.1000
1221 May 23a C	-0.410	- 14.6	+ 29.4	0.0500
1230 May 14 B,E	+0.062	- 5.8	- 1.9	0.1000
1230 May 14 E,G	-0.953	- 6.9	- 3.1	0.0783
1230 May 14 E,BN	+0.164	- 3.6	- 1.8	0.1414
1230 May 14 E,Sc	+0.128	- 5.4	- 2.5	0.1003
1232 Oct 15 E,G	-0.896	+ 4.3	+ 2.1	0.1111
1236 Aug 3 E,F	+2.383	+ 5.9	+ 1.9	0.1000
1236 Aug 3 E,Sc	+0.369	- 0.5	+ 9.1	0.1008
1239 Jun 3a B,E	-1.493	- 4.9	+ 12.1	0.1000
1239 Jun 3b B,E	-1.572	- 5.3	+ 12.5	0.1000
1239 Jun 3a E,F	-0.131	- 4.0	+ 9.2	0.1414
1239 Jun 3b E,F	+0.115	- 17.8	+ 40.5	0.0321
1239 Jun 3c E,F	-0.209	- 3.9	+ 10.5	0.1107
1239 Jun 3d E,F	-0.029	- 4.1	+ 11.7	0.1000
1239 Jun 3e E,F	-0.089	- 12.1	+ 29.1	0.0437

Designation	r_i	$n_{Y,i}$	$n_{N,i}$	$\sigma_{n,i}$
1239 Jun 3a E,G	+0.277	− 2.4	+ 12.0	0.0783
1239 Jun 3b E,G	−0.755	− 1.4	+ 9.1	0.1000
1239 Jun 3a E,I	−0.191	− 0.9	+ 6.4	0.1414
1239 Jun 3b E,I	−0.280	− 3.7	+ 20.7	0.0454
1239 Jun 3c E,I	−0.922	− 5.0	+ 29.0	0.0321
1239 Jun 3d E,I	+0.212	− 5.3	+ 29.6	0.0321
1239 Jun 3e E,I	−0.269	− 1.5	+ 9.1	0.1000
1239 Jun 3f E,I	−0.395	− 2.2	+ 13.4	0.0681
1239 Jun 3g E,I	−0.242	− 1.9	+ 9.6	0.1000
1239 Jun 3h E,I	−0.554	− 4.3	+ 22.0	0.0437
1239 Jun 3i E,I	−0.342	− 2.2	+ 9.9	0.1000
1239 Jun 3j E,I	−0.347	− 3.4	+ 14.7	0.0681
1239 Jun 3k E,I	+0.432	− 5.5	+ 23.1	0.0437
1239 Jun 3ℓ E,I	−0.428	− 4.6	+ 22.4	0.0437
1239 Jun 3m E,I	−0.520	− 5.9	+ 30.2	0.0321
1239 Jun 3n E,I	+0.394	− 0.8	+ 8.6	0.1000
1239 Jun 3o E,I	+0.175	− 1.4	+ 9.3	0.1000
1239 Jun 3p E,I	+0.545	− 4.4	+ 29.0	0.0321
1239 Jun 3q E,I	−0.206	− 2.8	+ 15.2	0.0618
1239 Jun 3r E,I	−0.090	− 1.2	+ 6.6	0.1414
1239 Jun 3s E,I	+0.230	− 1.9	+ 14.6	0.0618
1239 Jun 3a E,CE	−0.649	− 1.0	+ 8.6	0.1000
1239 Jun 3b E,CE	−0.692	− 0.7	+ 8.4	0.1000
1241 Oct 6a B,E	+0.656	+ 13.2	− 19.3	0.1000
1241 Oct 6b B,E	−0.364	+ 16.7	− 24.6	0.0783
1241 Oct 6c B,E	+0.750	+ 13.1	− 19.1	0.1000
1241 Oct 6d B,E	+0.495	+ 9.3	− 13.6	0.1414
1241 Oct 6a E,F	+0.273	+ 9.7	− 14.1	0.1414
1241 Oct 6b E,F	+0.362	+ 13.5	− 19.8	0.1000
1241 Oct 6a E,G	+0.278	+ 26.5	− 39.6	0.0490

Designation	$r_{\underline{i}}$	$\eta_{\underline{Y},\underline{i}}$	$\eta_{\underline{N},\underline{i}}$	$\sigma_{\eta,\underline{i}}$
1241 Oct 6b E,G	+0.126	+ 13.1	− 19.4	0.1000
1241 Oct 6c E,G	+0.047	+ 38.0	− 57.6	0.0321
1241 Oct 6d E,G	+0.153	+ 29.5	− 43.9	0.0437
1241 Oct 6e E,G	+0.071	+ 12.7	− 19.1	0.1000
1241 Oct 6f E,G	−0.259	+ 36.1	− 56.1	0.0321
1241 Oct 6g E,G	+0.103	+ 39.9	− 59.5	0.0321
1241 Oct 6h E,G	+0.242	+ 28.9	− 42.7	0.0454
1241 Oct 6i E,G	+0.343	+ 40.8	− 60.4	0.0321
1241 Oct 6k E,G	+0.350	+ 21.0	− 31.4	0.0618
1241 Oct 6ℓ E,G	−0.037	+ 12.1	− 18.5	0.1000
1241 Oct 6m E,G	−0.078	+ 27.9	− 42.6	0.0437
1241 Oct 6n E,G	−0.037	+ 38.9	− 59.2	0.0321
1241 Oct 6o E,G	−0.051	+ 27.8	− 41.6	0.0454
1241 Oct 6p E,G	−0.087	+ 28.8	− 43.2	0.0437
1241 Oct 6q E,G	+0.137	+ 13.2	− 19.5	0.1000
1241 Oct 6r E,G	+0.182	+ 13.2	− 19.6	0.1000
1241 Oct 6s E,G	+0.142	+ 9.2	− 13.6	0.1414
1241 Oct 6t E,G	−0.400	+ 27.0	− 41.9	0.0437
1241 Oct 6u E,G	+0.233	+ 29.1	− 43.7	0.0437
1241 Oct 6v E,G	+0.511	+ 20.9	− 30.7	0.0636
1241 Oct 6w E,G	−0.086	+ 19.7	− 29.9	0.0618
1241 Oct 6x E,G	−0.174	+ 38.9	− 59.2	0.0321
1241 Oct 6y E,G	+0.016	+ 29.1	− 43.5	0.0437
1241 Oct 6 E,I	+0.178	+ 13.4	− 19.8	0.1000
1241 Oct 6 E,BN	+0.414	+ 13.3	− 19.4	0.1000
1241 Oct 6a E,BN	+0.334	+ 13.1	− 19.2	0.1000
1241 Oct 6a E,CE	+0.037	+ 29.3	− 43.9	0.0437
1241 Oct 6b E,CE	−0.146	+ 28.8	− 43.5	0.0437
1241 Oct 6c E,CE	−0.171	+ 38.9	− 59.2	0.0321
1241 Oct 6d E,CE	−0.115	+ 39.3	− 59.2	0.0321

Designation	$r_{\underline{i}}$	$\eta_{\underline{Y},\underline{i}}$	$\eta_{\underline{N},\underline{i}}$	$\sigma_{\eta,\underline{i}}$
1241 Oct 6e E,CE	-0.400	+ 24.3	- 37.6	0.0490
1241 Oct 6f E,CE	-0.446	+ 27.9	- 42.6	0.0437
1241 Oct 6g E,CE	-0.501	+ 27.5	- 42.3	0.0437
1241 Oct 6h E,CE	-0.025	+ 12.8	- 19.2	0.1000
1241 Oct 6i E,CE	-0.066	+ 8.9	- 13.4	0.1414
1241 Oct 6 E,PR	-0.391	+ 10.9	- 17.4	0.1000
1241 Oct 6 E,Sc	-0.176	+ 7.6	- 12.1	0.1425
1245 Jul 25 E,G	+2.376	- 5.7	- 0.8	0.1000
1255 Dec 30 B,E	-1.806	- 2.7	- 5.2	0.1000
1255 Dec 30 E,CE	-1.366	- 6.1	- 0.9	0.1000
1261 Apr 1 B,E	-2.623	- 10.7	+ 19.4	0.1000
1261 Apr 1 E,G	-1.209	- 9.4	+ 15.8	0.1414
1261 Apr 1 E,CE	-1.366	- 13.8	+ 23.0	0.1000
1263 Aug 5a B,E	+1.168	+ 12.0	- 22.0	0.1000
1263 Aug 5b B,E	+1.141	+ 11.8	- 21.8	0.1000
1263 Aug 5c B,E	+1.099	+ 11.8	- 21.9	0.1000
1263 Aug 5 E,G	+0.529	+ 11.3	- 21.4	0.1000
1263 Aug 5a E,G	+0.289	+ 10.6	- 20.4	0.1003
1263 Aug 5b E,G	+0.296	+ 10.6	- 20.5	0.1000
1263 Aug 5c E,G	+0.353	+ 10.4	- 20.3	0.1000
1263 Aug 5a E,I	+0.695	+ 11.8	- 22.1	0.1000
1263 Aug 5b E,I	+0.845	+ 12.2	- 22.6	0.1000
1263 Aug 5 E,BN	+0.885	+ 11.6	- 21.8	0.1000
1263 Aug 5 E,CE	-0.720	+ 14.3	- 27.3	0.0783
1263 Aug 5a E,CE	+0.280	+ 10.7	- 20.7	0.1000
1263 Aug 5 E,PR	-0.035	+ 9.8	- 19.7	0.1000
1263 Aug 5a E,Sc	+0.241	+ 8.9	- 18.2	0.1000
1263 Aug 5b E,Sc	+0.312	+ 10.2	- 19.8	0.1000
1263 Aug 5c E,Sc	+0.139	+ 6.8	- 13.6	0.1414
1267 May 25 E,G	-1.633	- 6.7	+ 0.5	0.1000

Designation	r_i	$\eta_{Y,i}$	$\eta_{N,i}$	$\sigma_{\eta,i}$
1267 May 25a E,G	−1.969	− 5.7	− 0.5	0.1004
1267 May 25 E,CE	−1.568	− 6.5	+ 0.3	0.1000
1267 May 25a E,CE	−1.608	− 5.9	− 0.3	0.1000
1267 May 25b E,CE	−1.495	− 5.8	− 0.3	0.1000
1267 May 25c E,CE	−1.747	− 7.3	+ 1.2	0.1000
1267 May 25 M,B	−0.132	− 7.5	− 5.9	0.0454
1270 Mar 23 E,G	−0.459	− 1.6	− 3.4	0.1414
1270 Mar 23a E,CE	−0.245	− 2.8	− 4.0	0.1000
1270 Mar 23b E,CE	−0.238	− 2.9	− 4.1	0.1000
1270 Mar 23 E,Sc	−0.710	− 2.1	− 4.9	0.1000
1288 Apr 2 B,E	+3.610	− 2.2	− 3.7	0.1000
1288 Apr 2 B,W	+3.079	− 3.4	− 3.9	0.0788
1290 Sep 5a E,F	+0.839	− 0.4	+ 7.9	0.1000
1290 Sep 5b E,F	−0.263	− 0.5	+ 10.1	0.0783
1290 Sep 5 E,G	+0.058	+ 0.9	+ 6.2	0.1000
1290 Sep 5 E,I	−0.698	+ 0.9	+ 5.6	0.1111

COEFFICIENTS OF CONDITION FOR ECLIPSES
FROM 1310 TO 1399

Designation	r_i	$\eta_{Y,i}$	$\eta_{N,i}$	$\sigma_{n,i}$
1310 Jan 31a B,E	−0.192	− 6.7	+ 1.8	0.1000
1310 Jan 31b B,E	−0.349	− 6.4	+ 1.4	0.1000
1310 Jan 31a E,F	+0.148	− 7.3	+ 2.7	0.1000
1310 Jan 31b E,F	−1.140	− 9.5	+ 3.3	0.0783
1310 Jan 31c E,F	−1.140	− 9.5	+ 3.3	0.0783
1310 Jan 31a E,G	−0.866	− 9.1	+ 3.2	0.0783
1310 Jan 31b E,G	+0.989	− 11.9	+ 4.9	0.0636
1310 Jan 31a E,I	+0.538	− 4.2	+ 2.0	0.2000
1310 Jan 31b E,I	+0.573	− 4.2	+ 2.1	0.2000
1310 Jan 31 E,BN	+0.144	− 7.1	+ 2.4	0.1000
1310 Jan 31a E,CE	+0.924	− 7.6	+ 3.1	0.1000
1310 Jan 31b E,CE	+0.818	− 7.5	+ 3.1	0.1000
1310 Jan 31c E,CE	+0.796	− 7.7	+ 3.2	0.1000
1312 Jul 5 E,G	+1.865	− 8.0	+ 16.9	0.0783
1321 Jun 26a E,G	+0.245	− 15.1	+ 3.9	0.0437
1321 Jun 26c E,G	−0.785	− 14.0	+ 2.3	0.0437
1321 Jun 26d E,G	−0.212	− 6.2	+ 1.3	0.1000
1321 Jun 26 E,CE	+0.322	− 6.5	+ 1.6	0.1000
1321 Jun 26a E,PR	+0.478	− 9.4	+ 1.7	0.0636
1321 Jun 26b E,PR	−0.279	− 5.8	+ 0.7	0.1000
1330 Jul 16 B,E	+0.421	+ 6.2	− 1.4	0.1000
1330 Jul 16a E,I	−1.955	+ 5.9	− 1.9	0.1111
1330 Jul 16b E,I	+0.379	+ 4.7	− 1.4	0.1414
1330 Jul 16c E,I	−1.961	+ 6.0	− 1.8	0.1111
1330 Jul 16 E,Sc	−0.146	+ 3.6	+ 1.8	0.1011
1331 Nov 30 M,B	+2.267	+ 9.4	− 14.8	0.1000
1333 May 14 E,I	+0.851	+ 3.7	+ 3.7	0.0783
1337 Mar 3a M,B	−1.337	− 10.9	+ 19.2	0.1000
1337 Mar 3b M,B	−0.296	− 8.6	+ 15.0	0.1414
1339 Jul 7 B,E	−1.462	+ 14.4	− 28.7	0.0522

Designation	r_i	$n_{Y,i}$	$n_{N,i}$	$\sigma_{n,i}$
1339 Jul 7 E,I	−0.415	+ 11.7	− 21.5	0.0783
1339 Jul 7 E,CE	+0.114	+ 19.0	− 35.9	0.0437
1339 Jul 7 E,PR	−0.403	+ 7.7	− 15.0	0.1000
1339 Jul 7 E,Sc	−0.133	+ 3.1	− 9.3	0.1014
1341 Dec 9 M,B	+2.580	+ 2.3	+ 2.4	0.1000
1354 Sep 17 E,I	+0.081	+ 7.0	− 12.5	0.1107
1361 May 5 B,E	−3.356	− 5.4	+ 0.8	0.1000
1361 May 5 E,I	−1.951	− 6.6	+ 2.1	0.1000
1361 May 5 M,B	−0.285	− 7.8	+ 0.6	0.0618
1364 Mar 4 E,BN	+1.883	− 5.8	+ 2.5	0.1000
1384 Aug 17a B,E	−1.367	+ 4.9	− 1.2	0.1000
1384 Aug 17b B,E	−1.367	+ 4.9	− 1.2	0.1000
1384 Aug 17 E,I	+0.783	+ 9.4	− 4.4	0.0636
1384 Aug 17 E,BN	+0.354	+ 4.6	1.5	0.1111
1386 Jan 1a E,I	−0.322	+ 2.2	− 9.5	0.0681
1386 Jan 1b E,I	+1.069	+ 3.7	− 14.9	0.0437
1399 Oct 29 E,I	+0.987	+ 1.1	+ 1.6	0.1414

COEFFICIENTS OF CONDITION FOR ECLIPSES
FROM 1406 TO 1486

Designation	r_i	$\eta_{Y,i}$	$\eta_{N,i}$	$\sigma_{n,i}$
1406 Jun 16 B,E	−0.391	− 4.2	+ 9.1	0.1000
1406 Jun 16 E,G	+0.469	− 4.8	+ 9.9	0.1000
1406 Jun 16 E,I	−0.708	− 5.9	+ 11.8	0.0863
1408 Oct 19a E,I	−1.438	+ 5.2	− 9.5	0.1414
1408 Oct 19b E,I	−0.893	+ 6.7	− 12.1	0.1107
1408 Oct 19 E,CE	−2.800	+ 6.9	− 12.8	0.1000
1409 Apr 15a E,I	+1.025	− 0.2	+ 3.7	0.1414
1409 Apr 15b E,I	+0.365	− 0.3	+ 4.8	0.1107
1415 Jun 7a E,G	−0.112	− 13.4	+ 2.8	0.0321
1415 Jun 7b E,G	−0.037	− 4.3	+ 1.1	0.1000
1415 Jun 7a E,I	−0.791	− 6.0	+ 1.8	0.0783
1415 Jun 7b E,I	−0.410	− 6.1	+ 2.2	0.0783
1415 Jun 7c E,I	+0.128	− 4.5	+ 1.2	0.1000
1424 Jun 26 B,E	+0.836	+ 3.4	− 0.1	0.1000
1431 Feb 12a E,I	−0.140	− 3.7	+ 7.2	0.1414
1431 Feb 12b E,I	−0.198	− 5.3	+ 10.2	0.1000
1431 Feb 12c E,I	−0.040	− 5.3	+ 10.2	0.1000
1431 Feb 12d E,I	+0.165	− 5.5	+ 10.4	0.1000
1431 Feb 12e E,I	+0.081	− 16.8	+ 32.1	0.0321
1433 Jun 17 E,F	−1.706	+ 4.6	− 9.1	0.1090
1448 Aug 29a E,I	+1.358	+ 9.4	− 21.2	0.0636
1448 Aug 29b E,I	−0.033	+ 5.3	− 12.1	0.1107
1460 Jul 18 E,I	−0.303	− 2.6	+ 6.5	0.1107
1465 Sep 20a E,I	+0.186	0.0	− 6.1	0.0636
1465 Sep 20b E,I	+1.480	+ 0.1	− 4.0	0.1000
1473 Apr 27 Is,E	−0.568	− 8.2	+ 5.3	0.0437
1478 Jul 29 E,I	−1.246	+ 3.4	− 1.4	0.1000
1485 Mar 16 E,I	+0.814	− 1.9	+ 7.7	0.0636
1485 Mar 16 E,CE	+0.046	− 2.1	+ 10.1	0.0437
1486 Mar 6 E,CE	−0.077	+ 0.3	+ 3.6	0.0783

TABLE A.I.17

COEFFICIENTS OF CONDITION FOR ECLIPSES
FROM 1502 TO 1567

Designation			r_i	$\eta_{Y,i}$	$\eta_{N,i}$	$\sigma_{\eta,i}$
1502 Oct	1	E,CE	−0.335	+ 3.2	− 10.0	0.0783
1502 Oct	1	Is,E	+2.425	+ 5.0	− 12.3	0.1000
1513 Mar	7	Is,E	+0.479	− 2.3	+ 0.8	0.1000
1544 Jan	24	E,G	−0.087	− 2.5	+ 12.1	0.0321
1560 Aug	21	E,SP	+1.304	+ 42.0	−102.3	0.0069
1567 Apr	9	E,I	+0.016	− 25.4	+ 14.3	0.0063

TABLE A.I.18

VALUES OF THE PARAMETER y AS A FUNCTION OF TIME
DURING THE HISTORICAL PERIOD

Epoch	y	$\sigma(y)$
− 660	−24.4	2.3
− 551	−20.3	3.1
− 398	−22.5	2.9
− 166	−22.6	1.4
− 44	−18.4	2.4
122	−22.0	3.1
415	−21.2	4.2
602	−19.1	4.5
772	−38.1	7.9
878	−21.0	6.4
1005	−22.1	6.1
1128	−10.5	4.3
1174	+ 4.0	7.8
1248	−12.8	5.4
1354	−17.6	17.9
1446	−33.0	30.0
1534	+ 6.3	20.2

APPENDIX II

TWO DEFINITIONS OF THE MAGNITUDE OF AN ECLIPSE

We define the magnitude of an eclipse with the aid of
Figure A.II.1. At the risk of a suit for copyright in-
fringement, I have copied this figure from Chapter VI.7 of
the Syntaxis, even to the lettering.† The circle αβγδ
represents the object that is being eclipsed while the circle αζγη represents the object that is doing the eclipsing.
In the case of a solar eclipse, αβγδ is the sun and αζγη is
the moon. In the case of a lunar eclipse, αβγδ is the moon
and αζγη is the umbra of the earth's shadow at the distance
of the moon.

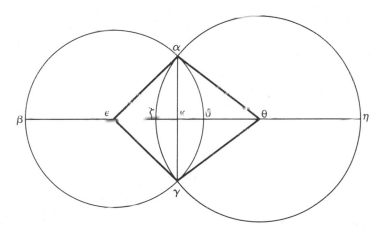

Figure A.II.1. The geometry of an eclipse. The circle αβγδ represents the disk that is being
eclipsed and the circle αζγη represents the disk that is doing the eclipsing. The line ζδ is
the part of the diameter of the circle αβγδ that is eclipsed and the lenticular region αζγδ
is the part of its area that is eclipsed. This figure is copied from Chapter VI.7 of the
Syntaxis, except that I have used lower case instead of upper case letters.

†However, I have used lower case letters instead of the
upper case ones that Ptolemy must have used. Lower case
Greek letters were not invented until long after the time
of Ptolemy.

The line $\overline{\zeta\delta}$ is the part of the diameter of the circle $\alpha\beta\gamma\delta$ that is eclipsed. Today, we define the magnitude of an eclipse, which I denote by m_D, as the ratio of $\overline{\zeta\delta}$ to the diameter $\overline{\beta\delta}$; I take the diameter as 2 by definition. Thus,

$$m_D = \overline{\zeta\delta}/2. \qquad\qquad (A.II.1)$$

I believe that astronomers at least since Regiomontanus have defined the magnitude in this way, although both Regiomontanus and Walther frequently take care to specify that they are referring to what we may call the magnitude of the diameter. This suggests that some other kind of magnitude had been in common use in earlier times.

In fact, Ptolemy [Syntaxis, Chapter VI.7) gives another definition of the magnitude of an eclipse: " . . most of those who observe eclipses do not measure the magnitude of the shadowing by the diameters of the disks but by their entire surfaces, comparing grossly by sight what they see of the disk with that which they do not see, . ." That is, most astronomers consider the area of the lenticular region $\alpha\zeta\gamma\delta$ and define the magnitude as the ratio of this area to the area of the circle $\alpha\beta\gamma\delta$; this area is π by the definition of the diameter $\overline{\beta\delta}$. I use the term "magnitude of the area" and the symbol m_A to denote the magnitude defined this way. Hence, we have

$$m_A = (\text{area } \alpha\zeta\gamma\delta)/\pi. \qquad\qquad (A.II.2)$$

The purpose of this appendix is to study the relation between m_D and m_A.

In deriving Equations A.II.1 and A.II.2, I have taken the smaller circle $\alpha\beta\gamma\delta$ in Figure A.II.1 to be the one that is being eclipsed. For slightly more than half of all solar eclipses, the sun is larger than the moon and the larger circle is the one that is eclipsed. For these eclipses, we should use the diameter and area, respectively, of circle $\alpha\zeta\gamma\eta$ in Equations A.II.1 and A.II.2, rather than those of circle $\alpha\beta\gamma\delta$.

However, the magnitude m_A of the area was never used in any stage of astronomy from which we have accurate measurements, and thus we do not need to treat it accurately. For simplicity, I take the two circles to be equal for all solar eclipses. For lunar eclipses, I calculate the ratio of the radii $\overline{\alpha\theta}$ and $\overline{\alpha\epsilon}$ using average values of the lunar and solar distances. From the data on pages 215 and 257 of the Explanatory Supplement [1977], we find that the ratio is 2.697 21 for the average distances. Specifically, the average radius of the moon is 15'.5430 and the average radius of the umbra is 41'.9228.

It is interesting to note that Ptolemy takes the ratio to be 31;12/12;0 = 2.6 for all lunar eclipses. He takes the ratio of the lunar to the solar radius to be unity when the moon is at its maximum distance, thus denying the possibility of annular solar eclipses. For the mean distances, he takes the ratio to be 12;20/12;0 = 1.028. Actually, as I

just said, the value should be slightly less than 1, but I
will use 1 for simplicity for solar eclipses.

Reverting to Figure A.II.1, we take the radius $\overline{\alpha\epsilon}$ to
be 1 and the radius $\overline{\alpha\Theta}$ to be $\underline{R}$, where $\underline{R}$ is either 1 or 2.697 21.
The distance $\overline{\zeta\delta}$ equals $2\underline{m}_D$ according to Equation A.II.1, so
the distance $\overline{\epsilon\Theta}$ equals $1 + \underline{R} - 2\underline{m}_D$. This gives us all the
sides in the triangle $\overline{\alpha\epsilon\Theta}$, and we can therefore solve for the
angles $\overline{\alpha\epsilon\kappa}$ and $\overline{\alpha\Theta\kappa}$, and for the altitude $\overline{\alpha\kappa}$. I will not take
the space to write out the formal solution.

The area of the sector $\overline{\alpha\epsilon\gamma\delta}$ is $(\pi/180°) \times \overline{\alpha\epsilon\kappa}$ if $\overline{\alpha\epsilon\kappa}$ is in
degrees, and the area of the sector $\overline{\alpha\Theta\gamma\zeta}$ is $(\pi\underline{R}^2/180°) \times \overline{\alpha\Theta\kappa}$
if the angle $\overline{\alpha\Theta\kappa}$ is also in degrees. The area of the
eclipsed region is the sum of these two sectors minus the
area of the quadrilateral $\overline{\alpha\epsilon\gamma\Theta}$, which equals the product of
its diagonals since the angle at κ is 90°. This gives us

$$\underline{m}_A = [(\pi/180°)(\overline{\alpha\epsilon\kappa} + \underline{R}^2 \times \overline{\alpha\Theta\kappa}) - (\overline{\alpha\gamma})(\overline{\epsilon\Theta})]/\pi.$$
$$(A.II.3)$$

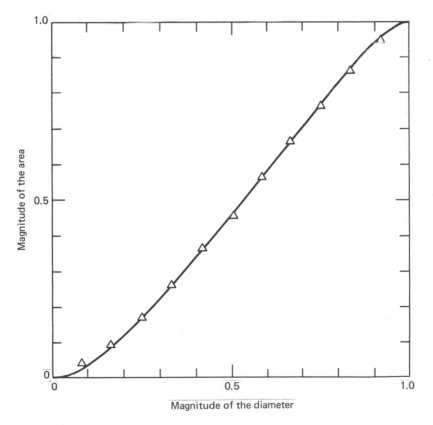

Figure A.II.2. The magnitude of the area as a function of the magnitude of the diameter
for a lunar eclipse. The points enclosed by triangles are the points tabulated by Ptolemy
in Chapter VI.8 of the *Syntaxis*.

Equation A.II.3 gives us $\underline{m}_A$ as a function of $\underline{m}_D$, and we could write it out in closed form if we wished to take the space. I have not tried to see if there is a closed form for $\underline{m}_D$ as a function of $\underline{m}_A$, but I suspect there is not. This may be the reason why Ptolemy tabulates $\underline{m}_A$ as a function of $\underline{m}_D$ but not vice versa.

Figure A.II.2 shows $\underline{m}_A$ as a function of $\underline{m}_D$ for lunar eclipses ($\underline{R}$ = 2.697 21), and Figure A.II.3 shows the same function for solar eclipses ($\underline{R}$ = 1). The triangles in the figures show the values tabulated by Ptolemy in Chapter VI.8 of the Syntaxis. We see that Ptolemy's values are reasonably accurate. The largest error in his $\underline{m}_A$ is about 0.02.†

Reverting to Figure A.II.1 for a moment, we see that all arcs αζγ that pass through the point ζ give the same value of $\underline{m}_D$, regardless of the value of $\underline{R}$. However, as $\underline{R}$ increases, the arc αζγ becomes straighter, and the area eclipsed becomes greater. In other words, for a given value of $\underline{m}_D$, $\underline{m}_A$ is a monotone increasing function of $\underline{R}$, and hence $\underline{m}_A$ is larger for lunar eclipses than for solar eclipses for all values of $\underline{m}_D$. We can see this from a comparison of Figures A.II.2 and A.II.3.

Further, in the limit $\underline{R}$ = ∞, the arc αζγ becomes a straight line and $\underline{m}_A$ becomes an antisymmetric function of $\underline{m}_D$ about the value 0.5. The greatest departure from antisymmetry occurs when R = 1, which applies for solar eclipses (Figure A.II.3). From this standpoint, the value R = 2.697 21 (lunar eclipses, Figure A.II.2) is more than halfway from 1 to ∞.

My programs for calculating the circumstances of eclipses are arranged to give the magnitude $\underline{m}_D$. When a record gives a value of $\underline{m}_A$ rather than of $\underline{m}_D$, I use Figures A.II.2 and A.II.3 to find the equivalent value of $\underline{m}_D$.

---

†Ptolemy actually expresses $\underline{m}_D$ and $\underline{m}_A$ in digits. To get a magnitude in digits, we multiply the axes in Figures A.II.2 and A.II.3 by 12. The digit was used to express the magnitude of an eclipse well into modern times.

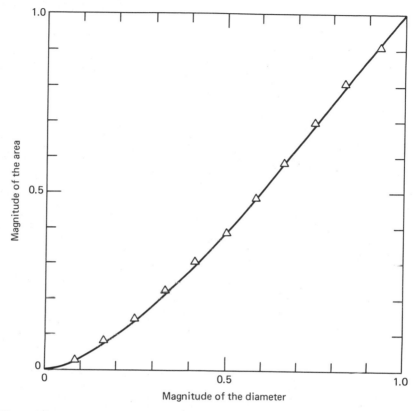

Figure A.II.3. The magnitude of the area as a function of the magnitude of the diameter for a solar eclipse. The points enclosed by triangles are the points tabulated by Ptolemy in Chapter VI.8 of the *Syntaxis*.

COORDINATES OF OBSERVING SITES

In this appendix, I give the latitude and longitude of the places where observations were made. The island of Hveen, where Tycho made most of his observations, is the only site for which I need an altitude, and I do not need it accurately. I use 50 meters, on the basis that I described in Section II.6.

The latitudes and longitudes that are used in the analysis of the observations are listed in Table A.III.1 on the next page. In preparing the table, I first looked up the coordinates in the tables of Times Atlas [1955]. As a precaution, I then measured the coordinates from the appropriate maps in the atlas and compared the measured values with the tabulated values. I found good agreement for all places except Antioch (Antakya), Turkey. For Antioch, the latitude on the map agrees well with that in the table, but the longitude from the map is 36° 06' while the table gives 36° 10'. I thank my colleague P. M. Schwimmer, of the Defense Mapping Agency, for supplying the coordinates of Antioch that appear in Table A.III.1. We see that the table in the Times Atlas is correct and that the map is wrong.

The coordinates of Hveen are those given in Section II.6. They apply specifically to the great mural quadrant, and they were obtained by a special survey of the site made in 1903/1904.

The observations from China cannot be assigned to a specific place. The best we can do is to assume that they could have been made anywhere within a rectangle that is luckily much smaller than modern China. The table gives the extremes of latitude and longitude at the corners of the rectangle.

THE COORDINATES OF PLACES USED IN THE ANALYSIS

Place	Latitude °	Longitude °
Alexandria	31.18	29.84
Antakya,[a] Turkey	36.20	36.16
Babylon	32.55	44.42
Baghdad	33.33	44.43
Cairo	30.05	31.25
China	32.91/35.21[b]	109.94/117.08[b]
Durham, England	54.78	− 1.57
Ghazni, Afghanistan	33.55	68.47
Hveen[c]	55.90685	12.69861
Kabul, Afghanistan	34.50	69.17
(Great) Malvern, England	52.12	− 2.32
Melk, Austria	48.23	15.35
Nürnberg, Germany	49.45	11.08
Orange, France	44.13	4.80
Padua, Italy	45.40	11.88
Raqqa, Syria	35.95	39.05
Rome	41.88	12.50
Samarkand, USSR	39.67	66.95
Toledo, Spain	39.87	− 4.03
Ujjain[d], India	23.18	75.83
Vienna	48.22	16.37
Viterbo, Italy	42.40	12.10

[a]Antioch.

[b]The limits in latitude and longitude, respectively.

[c]The great mural quadrant.

[d]Formerly called Arin.

APPENDIX IV

THE LIBRARY AT ALEXANDRIA

The library at Alexandria is the ancient library that is best known to modern readers, even though Alexandria is fairly new as ancient cities go. Alexander the Great founded the city in -331 on the site of some fishing villages, with the intention that it should be the link between Greece and the Nile valley. After his death in -322, his generals fought each other for the succession and ultimately divided his empire among themselves.

Egypt fell to the portion of the general later known as the king Ptolemy I or Ptolemy Soter, who was the Egyptian ruler de facto from -322, and the ruler de jure from -305, until he resigned in favor of his son in -284. In spite of his preoccupation with the wars of the times, Ptolemy Soter was a strong supporter of learning. He founded a library and museum in Alexandria, he made the city an important center of scholarship, and he himself wrote a biography of Alexander that has received considerable praise.

Ancient estimates of the size of the library put it in the neighborhood of half a million books. However, a book did not mean then what it does now. For example, Ptolemy's Syntaxis [Ptolemy, ca. 142],† which is not particularly large as modern books go, is divided into thirteen parts that were each called books in ancient usage. It is reasonable to say that an ancient book contained about a tenth of the material of a modern book, so that the library at Alexandria contained about 50 000 books in modern terms. This is still large in terms of the facilities of the time.

According to many standard histories, the library at Alexandria was burned twice, at times almost seven centuries apart. The story of the second burning is easier to discuss, so let us turn to it first. This burning, still according to the histories, occurred this way: After the Muslim conquest of Alexandria in 640, the Muslim general wrote to ask the caliph what to do with the library he had found there. The answer was that any book containing anything contrary to the Koran must be destroyed. Further, any book agreeing with the Koran was superfluous and should also be destroyed. Thus any book except the Koran was to be destroyed, regardless of its contents. In order to fulfill this order, the general had the books used as fuel for the public baths until they were all burned, which took six months.

---

†There is no reason to think that the astronomer Ptolemy was any relation to the royal Ptolemies. It is just a confusing identity of names.

This story first appears in any known source about six centuries after the event. I think we can safely dismiss it as a late piece of fiction. The first burning is a bit harder to judge.

It occurred, again according to the histories, in connection with the campaign of Julius Caesar in Alexandria in -47/-46. Plutarch [ca. 100] reports three difficulties Caesar had during this campaign: "The first difficulty he met with was want of water, for the enemies had turned the canals.† Another was, when the enemy endeavored to cut off his communication by sea, he was forced to divert that danger by setting fire to his own ships, which, after burning the docks, thence spread on and destroyed the great library. A third was, when in an engagement near Pharos,‡ he leaped from the mole into a small boat, to assist his soldiers who were in danger, and when the Egyptians pressed him on every side, he threw himself into the sea, and with much difficulty swam off. This was the time when, according to the story, he had a number of manuscripts in his hand, which, though he was continually darted at, and forced to keep his head often under water, yet he did not let go, but held them up safe from wetting in one hand, whilst he swam with the other."

Of these three difficulties, the first is credible and in fact corresponds to good tactics on the part of the Alexandrians. The other two are not credible. Destroying one's sea communications is not a way to preserve them. Caesar had his headquarters on land, and if he were working on his writing in Alexandria, it would have been there. He certainly would not have carried his manuscripts into a battle. These latter two "difficulties" are clearly two of many legends that grew up around the person of Caesar after his assassination.‡ Still, we should ask if there are kernels of truth behind the legends.

To answer this, we turn to the two contemporaneous accounts of the Alexandrian campaign that have survived. One is by Caesar himself [Caesar, ca. -43] and the other is by his close friend and partisan Hirtius [ca. -43].* These

---

†The canals that supplied water to the occupied part of Alexandria. I have seen this event described in these words: Ganymede cut off Caesar's water. Ganymede was one of the Alexandrian leaders.

‡An island in the harbor of Alexandria, where Caesar had his headquarters in the early part of the campaign.

‡Another example is the legend that there was a total eclipse of the sun when Caesar crossed the Rubicon. See Newton [1970, pp. 70ff].

*As I note in the references, some scholars believe that Hirtius was not the author of this reference. In citing it by Hirtius's name, I am not taking sides on this question. I merely use the name of Hirtius for convenience in citation.

sources describe different aspects of the campaign. Hirtius [Chapter XXI] describes an occasion on which Caesar was indeed forced to swim from a ship that was being attacked by the Alexandrians, but he does not even suggest the presence of any manuscripts. Caesar [Book III, Chapter 111] says that he burned all the Alexandrian ships, including those at the docks, but he does not mention any damage to any buildings on land, not even to the docks.† Still we can see how burning ships at the docks might be transmuted by legend into the burning of the docks themselves.

Thus neither contemporaneous source mentions the burning of any building in Alexandria, much less the burning of the library. Further, Andrieu [1954, pp. XLIIIff] lists and compares many accounts of the Alexandrian campaign that appeared within about three centuries of the event. I have read all the accounts he lists that are earlier than Plutarch, and I have found no mention that the library was burned during the campaign. The writing of Strabo is particularly instructive.

Strabo was a wealthy Greek who devoted his life to scholarly pursuits. He was born in about -63 or -62, and there is evidence that he traveled to Rome in -43, the year Caesar was assassinated. About 20 years after Caesar's Alexandrian campaign, he moved to Alexandria and lived there for at least five years. His two major works are his Historical Memoirs, which are lost except for fragments, and his Geography, which has survived almost intact. His Geography [Strabo, ca. 0] is one of the most important works on the subject that has survived from antiquity. In it, Strabo describes not only the physical circumstances of a place or region but also gives considerable information about its history and politics.

In Book XVII, Chapter I, Sections 5 - 13, he gives a detailed description of Alexandria and its history. He does not mention the library separately, but he says quite a bit about the great museum there, and the library was a part of

---

†Caesar numbers the Alexandrian ships at 72, not counting an unspecified number at the docks. In spite of his claim that he burned them all, it is unlikely that he burned many of them, and hence it is unlikely that he started a major conflagration that might have destroyed the library. Altogether he had to do with an Alexandrian fleet three times. The first is the time just mentioned, when he allegedly burned all the enemy ships. The second was a few days later, when he had a major engagement with the (already destroyed) Alexandrian fleet [Hirtius, Chapters 10 and 11] and did it serious damage, but was prevented by nightfall from destroying it completely. Following this [Hirtius, Chapter 13], the Alexandrians collected the custom ships at the mouths of the Nile, and some old ships that had been in the docks for a long time, and fitted them out for combat. They added these to the warships they still had and fought another naval battle. This was the time Caesar had to swim for it.

the museum. He does not mention any recent damage† to any
part of the museum.

Strabo's silence about the burning of the library is a
different order of silence from that of Caesar and Hirtius.
These two authors were writing a mixture of military history
and political propaganda, and their silence about the library, one could argue, simply means that they did not consider
its burning to be important for their purposes. Strabo, on
the other hand, was a dedicated scholar whose writing on
history and geography was founded on research done with
books. He lived in Alexandria for at least five years, and
it is safe to conclude that he did much historical and geographical research there. He was well acquainted with the
museum (he mentions that the museum had a refectory where
the scholars working there could take their meals) and he
undoubtedly talked to many people who were already mature
scholars during the time of Caesar's campaign. The burning
of the library would have been a catastrophe to these scholars because it would have caused disastrous interruption to
their studies. Thus the burning of the library would have
been a matter of great interest to Strabo. I find it incred-
ible that he would have been silent on the subject if the
library had been seriously damaged such a short time before.

Thus Strabo's failure to mention the burning of the
library is more than just the omission of an unimportant
point. It is almost an explicit statement that there was no
burning to mention. After all, we should remember that an
historian or chronicler records what happened, and we do not
expect him to outline all the things that did not happen.
In a history of Washington, D.C., for example, we would be
surprised to find a statement that the Library of Congress
was not destroyed twenty years ago, even though the statement would be true. But if the library had been destroyed
in 1960, we would be surprised to find this fact omitted
from a scholarly history written in 1980.

Thus we may conclude, well beyond a reasonable doubt,
that the library at Alexandria was not burned during Caesar's campaign there. The story that it was burned seems to
have originated about 1 $\frac{1}{2}$ centuries after the event, and we
may safely put it down as another of the many ancient legends about Caesar. Certainly no one should present the
burning as an established historical fact.

But if the library had not been burned, and was still
in existence in Plutarch's time, how could he write that it
had been destroyed? The answer comes from other ancient
sources, but still much later than Caesar, which say that
the library was rebuilt and restocked after the conquest of

---

†The reader should remember that he took up residence in
  Alexandria little more than twenty years after Caesar's
  campaign there.

Alexandria. These sources were apparently trying to reconcile the existence of the library with the legend that it
had been destroyed.

If the library did not burn, what happened to it? Probably the answer is that no human institution is immortal.
The first few of the royal Ptolemies were at once effective
rulers and strong supporters of learning. Few later rulers
of Alexandria, including the Romans, were both, and many
were neither. Further, the great period of Greek scholarship was pretty well over by about the year -200, and most
Greek scholarship after this time was commentary, not original research. Under these conditions, the library probably
continued to receive its routine annual appropriations for a
long time, but, with neither the direct personal interest of
the monarch nor the continued stimulus of fresh original
research, it would tend to become an ingrown and ineffective
institution, a center for pedantry rather than serious scholarship. We would expect its history after -200 to be one of
slow decay, with its collection and organization gradually
vanishing through neglect, but with it, nonetheless, remaining a major center for some centuries. Finally, there would
come a time when all the books had been carried away and the
building had crumbled.

SOME BABYLONIAN OBSERVATIONS THAT HAVE BEEN STUDIED BEFORE

Table IV.3 of an earlier work [Newton, 1976] lists a number of records of lunar synodic phenomena, including some lunar and solar eclipses. I have already given the analysis of the eclipses in Sections VIII.4 and VIII.5 of this volume.

In my earlier discussions of the other records, I decided that some of them were probably calculated and not observed. For example, there is a list of the times between moonset or moonrise and sunset or sunrise for a number of months in the year 73 of the Seleucid Era, with no accompanying remarks to indicate that the times were observed. Since we know the Babylonians had elaborate methods for calculating these times by this stage of history, the conservative course is to not use these records.

Table A.V.1 summarizes the records which indicate strongly that the times were observed. I discussed the sources in which these records were published in the earlier work (Section IV.4) and I will not repeat this information here, nor will I repeat the Babylonian dates of the observations. The first column in the table gives the date of the observation in the Julian calendar. The second column describes the nature of the interval that was measured. For example, the notation "MS − SR" means the time of moonset minus the time of sunrise; I believe that all the other entries in this column should now be obvious.

TABLE A.V.1

SOME BABYLONIAN MEASUREMENTS OF THE INTERVAL BETWEEN
SUNRISE OR SUNSET AND MOONRISE AND MOONSET

Date	Nature of interval	Measured length of interval $\underline{h}$	Calculated length of interval[a] $\underline{h}$	$10^4 \times \partial \underline{T}/\partial \underline{y}$	$10^4 \times \partial \underline{T}/\partial \underline{\dot{n}}_M$
−567 Apr 22	MS − SS	0.933	1.038	−120	+ 72
−567 May 6	MS − SR	0.267	0.240	− 50	+ 44
−567 May 18	SR − MR	1.533	1.550	+ 60	− 40
−567 Jun 20	MS − SS	1.333	1.501	−104	+ 64
−567 Jul 5	MS − SR	0.500	0.530	− 88	+ 72

[a]Calculated using $\underline{\dot{n}}_M = -28$ and $\underline{y} = -20$.

Date	Nature of interval	Measured length of interval h	Calculated length of interval[a] h	$10^4 \times \partial \underline{T}/\partial \underline{y}$	$10^4 \times \partial \underline{T}/\partial \underline{\dot{n}}_M$
-566 Feb 12	MS - SS	0.967	1.123	-102	+ 75
-566 Feb 27	MS - SR	1.133	1.464	- 52	+ 40
-566 Mar 14	MS - SS	1.667	1.703	-108	+ 72
-566 Mar 26	MS - SR	0.100	0.026	- 54	+ 40
-378 Oct 27	MS - SS	0.967	0.958	- 62	+ 46
-378 Nov 9	SS - MR	0.633	0.431	+ 42	- 40
-378 Nov 10	MS - SR	0.300	0.162	- 84	+ 66
-273 Oct 6	MS - SS	1.000	0.919	- 54	+ 34
-273 Oct 19	SS - MR	0.733	0.551	+ 44	- 32
-273 Oct 21	MS - SR	0.333	0.228	- 74	+ 58
-273 Nov 2	SR - MR	1.600	1.934	+ 84	- 58
-273 Nov 19	MR - SS	0.367	0.356	+ 64	- 80
-272 Mar 29	SR - MR	1.000	0.968	+ 46	- 32
-232 Oct 3	MS - SS	1.033	1.160	- 36	+ 32
-232 Oct 15	SS - MR	0.378	0.282	+ 46	- 30
-232 Oct 16	MR - SS	0.256	0.335	- 46	+ 32
-232 Nov 14	SS - MR	0.167	0.112	+ 52	- 38
-232 Nov 15	MS - SR	0.444	0.287	- 88	+ 56
-232 Nov 29	SR - MR	0.733	1.011	+ 70	- 56
-232 Dec 13	SS - MR	0.833	0.752	+ 58	- 44
-232 Dec 14	SR - MS	0.111	0.199	+ 86	- 54
-232 Dec 15	MS - SR	0.967	0.910	- 80	+ 52

[a]Calculated using $\underline{\dot{n}}_M = -28$ and $\underline{y} = -20$.

The third column gives the length of the measured interval, translated into modern hours, while the fourth
column gives the length of the interval that I calculate.
In the calculation, I assume that a rising or a setting of
either the sun or the moon occurs when the center of the
body is 50' below the true horizon. This is a modern definition, which says that a rising or setting is considered to
occur when the upper limb of the body is just on the apparent horizon, and it includes refraction and an allowance
for the average radius of the body. We do not know that the
Babylonians used this definition, although it seems plausible that they should have done so. If they used some other
definition, the error we make by using the modern definition
should be mostly cancelled when we consider all combinations
of risings and settings.

Also, in calculating the fourth column, I take $\underline{\dot{n}}_M = -28$
and $\underline{y} = -20$, in the customary units. The last two columns
give the derivatives of the measured interval $\underline{T}$ with respect

to $\underline{y}$ and $\underline{\dot{n}}_M$, respectively, multiplied by 10^4 for convenience. The derivatives were found by calculating the values of $\underline{T}$ for other values of $\underline{y}$ and $\underline{\dot{n}}_M$ and differentiating numerically.

As I have done in the rest of this work, I assume that the value $\underline{\dot{n}}_M = -28$ is correct, and I estimate only the value of $\underline{y}$ from the data. Before I do this, however, I wish to consider the question of whether we should weight the observations uniformly. I alluded briefly to this matter in Section VIII.2, where I decided to use equal weights for the data considered there. Here I wish to look into the matter in a little more detail.

I believe that I have seen a study somewhere, one that I cannot locate at the time of writing, in which the writer found that the error in Babylonian measurements of time intervals varied systematically with the length of the interval. Also, in Section II.3, I summarized a study I had made of the errors in Tycho's measurements of time intervals. I found that the errors were dominated by a random walk with a standard deviation of about 15 seconds for an interval of 10 minutes. For such a law of error, the error is proportional to the square root of the interval, and the weight should be inversely proportional to the first power of the interval.

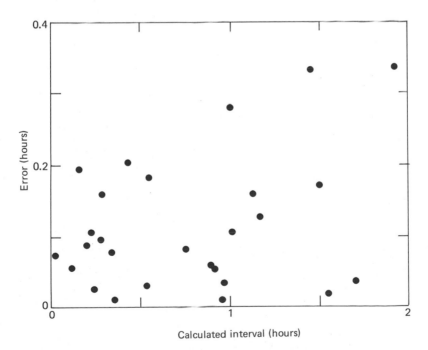

Figure A.V.1. The errors in Babylonian measurements of rising and setting times. The abscissa is the calculated (absolute) value of the interval between the rising or setting of the moon and the rising or setting of the sun. The ordinate is the absolute value of the error made in the corresponding Babylonian measurement. Perhaps surprisingly, the error does not seem to depend upon the length of the interval.

TABLE A.V.2

VALUES OF y FROM THE DATA IN TABLE A.V.1

Time span	$\underline{y}$	$\sigma(\underline{y})$
-567/-566	-10.5	5.8
-378	-22.9	13.3
-273/-232	-29.0	6.0

In order to see how the error depends upon the interval for the data in Table A.V.1, I have plotted the absolute value of the error† against the length of the interval in Figure A.V.1. The figure suggests that the three largest errors may not be members of the same population as the others, but it does not suggest any dependence of the error upon the length of the interval. Instead, the distribution of the errors seems to be independent of the length of the interval. Hence, I weight all the observations uniformly.

When we infer $\underline{y}$ from the data in Table A.V.1, we want to divide the observations into "time bins" in order that we may follow the time dependence of $\underline{y}$. In doing this, we naturally use the bins that are implicitly defined by the various tables in Appendix I. We want to put the observations from the years -567/-566 (these are from a single Babylonian year) into one group, those from -378 into another, and those from the years -273, -272, and -232 into a third. This gives the results shown in Table A.V.2.

In the table, the first columnn gives the span of years for the data in a group, the second column gives the value of $\underline{y}$ that we infer for that group, and the third column gives the standard deviation $\sigma(\underline{y})$ to be attached to the value of $\underline{y}$. In finding these standard deviations, we need the standard deviation of the errors in the observations. In finding this standard deviation, I assumed that it is the same for all groups. This assumption is reasonable, and without it we could not get a meaningful standard deviation for the second group.

Table IV.4 of the earlier work [Newton, 1976] lists a number of conjunctions of the moon with various stars, which I analyzed in Section X.5 of that work. Here, I will use most of the observations I used before, but I will omit some. Most of the omitted ones come from records which say

---

†Since the values used to calculate the fourth column in Table A.V.1 are close to the correct ones, it is adequate to take the error of a measurement to be the difference between the third and fourth columns in this study.

only that the moon was above or below the star in question. In the earlier work, I took such a statement to mean that the longitudes of the moon and the reference star were the same. I realize now that these terms imply nothing about the longitude, even when they are used alone.

TABLE A.V.3

SOME USAGES OF "BACK" AND "FORWARD"

Date	"Back" or "forward"	Direction shown by calculation
-273 Oct 18	back	- - -
-273 Oct 25	forward	- - -
-273 Oct 28	back	east
-250 Mar 6	back	east
-232 Dec 8	forward	- - -
-232 Dec 13	back	- - -
-232 Dec 17	back	west

In Section V.5 of the earlier work, I pointed out a number of recording problems, such as listing the wrong name for a star. In particular, there seems to be something systematically wrong with the records involving the star that we have interpreted to be η Tauri, and I can find no star that is consistent with the records. I have omitted observations that show recording problems without comment.

As I pointed out on page 120 of the earlier work, the terms used to indicate the east-west position of the moon relative to a star are unambiguous in most of the records. However, different terms are used in some (but not all) of the records which also say that the moon was over or under the reference star. When the east-west direction is given in such records, it is given by means of different terms that I have rendered as "back" and "forward". We also encountered a use of the term "back" in Section VII.4.

Altogether I have noticed seven records that use the terms "back" and "forward", and they are all records which also say whether the moon was over or under the reference star. These records are summarized in Table A.V.3. In the table, the first column gives the date of the observation, and the second column says whether the term used was "back" or "forward".

To explain the third column, let me give two examples. On -273 October 18, the moon was "back" of η Piscium by ½ ammat, or 1°.25. The longitude of η Piscium was 355°.245. Therefore the longitude of the moon was either 353°.995 or 356°.495, depending upon the meaning of "back". The calculated longitude of the moon, using any likely combination of the parameters $\underline{n}_M$ and $\underline{y}$, lies between these two values, and we cannot tell which interpretation is correct. I have indicated this situation by putting dashes in the last column.

On the other hand, on -273 October 28, the moon was "back" of α Leonis by 8 ubanu, or 0°.667. The longitude of α Leonis was 118°.391. Therefore the longitude of the moon was either 117°.724 or 119°.058. For any likely combination of the parameters, the calculated longitude of the moon is greater than 119°.058. Thus we may safely assume that the moon was east of the star.

Altogether, we can tell the direction of the moon from the star without ambiguity for three of the seven records in Table A.V.3. As it happens, the term involved was "back" for all three. The moon was east of the star twice and west of it once. Therefore, we cannot tell the meanings of "back" and "forward" on the basis of the available data.

For the four cases in Table A.V.3 in which we cannot tell the direction of the moon, the separation of the moon and the star is small and the cases are evenly divided between "back" and "forward". For this reason, we would not seriously affect the values we infer for the parameters if we used these records, and I did so in the earlier work [Newton, 1976, Section X.5]. There I took "back" to mean "east". On further reflection, I believe it is better not to use these four records. For one thing, by using them, I may have made it appear that we know the meanings of "back" and "forward" when we do not.

While I will not use these four records, I will use the three for which we can settle the direction of the moon by calculation.

As I pointed out on page 355 of the earlier work, the errors follow a normal pattern out to errors of about 2°, but there are many more errors larger than this than we expect. These are probably errors made in writing down the magnitude of the separation between the moon and the reference star, and I have omitted all records showing an error greater than 2°.

SOME BABYLONIAN MEASUREMENTS OF THE LONGITUDE OF THE
MOON REFERRED TO THE STARS

Date	Hour[a]	Measured longitude of the moon °	Calculated longitude of the moon[b] °	$10^4 \times \partial\lambda/\partial\underline{y}$	$10^4 \times \partial\lambda/\partial\underline{\dot{n}}_M$
-567 Apr 29	16.305	138.943	138.599	-1310	+ 906
-378 Nov 9	14.950	38.391	38.232	-1030	+ 768
-378 Nov 15	2.793	105.990	105.505	-1130	+ 776
-378 Nov 16	2.810	120.699	118.809	-1152	+ 776
-378 Nov 20	2.875	174.998	174.995	-1254	+ 774
-273 Oct 8	15.503	232.307	232.463	-1058	+ 700
-273 Oct 13	15.400	295.604	295.904	- 932	+ 704
-273 Oct 24	2.455	65.001	64.438	-1050	+ 706
-273 Oct 28	2.514	119.058	120.054	-1106	+ 708
-273 Oct 30	2.544	148.243	148.776	-1124	+ 708
-273 Oct 31	2.560	162.620	163.273	-1128	+ 706
-273 Nov 1	2.575	177.279	177.790	-1126	+ 704
-273 Nov 2	2.590	191.036	192.253	-1120	+ 700
-273 Nov 21	2.894	73.742	74.175	-1082	+ 704
-232 Oct 8	15.481	268.163	268.432	- 948	+ 678
-232 Oct 14	15.358	352.060	352.505	-1140	+ 678
-232 Oct 18	2.383	45.409	43.972	-1128	+ 680
-232 Oct 19	2.397	59.295	58.916	-1100	+ 682
-232 Oct 22	2.440	100.510	101.091	- 996	+ 684
-232 Oct 26	2.498	152.317	151.762	- 908	+ 680
-232 Nov 4	14.992	264.162	264.943	- 934	+ 676
-232 Nov 14	14.869	44.993	46.159	-1122	+ 674
-232 Nov 27	3.005	208.415	208.284	- 894	+ 676
-232 Nov 28	3.021	221.248	220.294	- 900	+ 674
-232 Dec 4	14.755	301.168	300.341	- 994	+ 678
-232 Dec 9	14.757	10.357	10.319	-1092	+ 682
-232 Dec 11	14.761	41.243	39.774	-1102	+ 680
-232 Dec 15	3.262	88.817	89.282	-1056	+ 676
-232 Dec 17	3.285	117.710	116.945	-1004	+ 680
-232 Dec 20	3.317	156.936	155.823	- 930	+ 682
-232 Dec 24	3.354	204.211	204.328	- 894	+ 680
-232 Dec 25	3.363	217.169	216.329	- 898	+ 678
-232 Dec 26	3.370	227.875	228.423	- 908	+ 678
-231 Jan 4	14.950	351.640	352.290	-1064	+ 682
-231 Jan 7	14.987	35.909	35.370	-1076	+ 684

[a] In Greenwich mean time.

[b] Calculated using $\underline{\dot{n}}_M = -28$ and $\underline{y} = -20$.

Date	Hour[a]	Measured longitude of the moon °	Calculated longitude of the moon[b] °	$10^4 \times \partial\lambda/\partial\underline{y}$	$10^4 \times \partial\lambda/\partial\underline{\dot{n}}_M$
-231 Jan 8	15.000	48.103	49.755	-1070	+ 682
-231 Jan 13	3.429	109.795	111.452	-1008	+ 676
-231 Feb 3	15.376	31.241	31.952	-1074	+ 682
-231 Mar 1	15.724	14.103	12.915	-1112	+ 678
-231 Mar 3	15.748	42.489	42.462	-1086	+ 682
-231 Mar 5	15.771	71.234	71.067	-1044	+ 684
-231 Mar 16	2.541	203.429[c]	203.429[c]	- 890	+ 676

[a]In Greenwich mean time.

[b]Calculated using $\underline{\dot{n}}_M = -28$ and $\underline{y} = -20$.

[c]There is no misprint here. These values are identical to the number of significant figures shown.

The remaining records are summarized in Table A.V.4. The first column of the table gives the date of an observation and the second column gives the hour. As I decided in Section VII.4, I take the hour to be 45 minutes after sunset or before sunrise, whichever is appropriate.† The third column gives the longitude of the moon that we infer from the record, and the fourth column gives the longitude that we calculate using $\underline{\dot{n}}_M = -28$ and $\underline{y} = -20$. Finally, the last two columns give the partial derivatives of the calculated longitude with respect to the parameters indicated.

For any given date, it is a simple matter to calculate the partial derivatives, and I did not even tabulate them in Table VIII.2. However, all the observations in Table VIII.2 were made within a single Babylonian year, and it was satisfactory to take the partial derivatives to be the same for all the observations. Here, the time span is too large to allow this simplification, and I have used the values of the partial derivatives that are listed separately for each observation.

---

†The hour is, of course, based upon calculation. The hour depends somewhat upon the values used for $\underline{\dot{n}}_M$ and $\underline{y}$, but within the range of uncertainty of these parameters, the dependence of the hour upon the parameters is only a few seconds. This is negligible.

TABLE A.V.5

VALUES OF $\underline{y}$ FROM THE DATA IN TABLE A.V.4

Time span	$\underline{y}$	$\sigma(\underline{y})$
-567	-22.6	6.4
-378	-25.5	3.6
-273/-231	-19.1	1.3

As before, we want to divide the observations into time bins. We must put the single observation from -567 into one bin, the four from -378 into another, and all those from -273 to -231 into a final one. The resulting values of $\underline{y}$ are shown in Table A.V.5, which has the same format as Table A.V.2. The standard deviation in the measured longitudes is almost exactly 50 minutes of arc.

The results from Tables A.V.2 and A.V.5 will be combined with other results in Section X.1.

APPENDIX VI

SOME PROPERTIES OF THE MOMENT OF INERTIA OF THE OCEANS

In this appendix, I calculate the change in the moment of the inertia of the earth that results from two types of change in the oceans. In the first calculation, I see what happens if the upper layer of the oceans changes temperature. In the second one, I see what happens if some continental ice melts and runs into the oceans.

We are only interested in approximate values of the changes, so I can make some simple assumptions. I assume that the oceans cover the entire earth (in spite of postulating continental ice), and I assume that the density of ocean water is nominally 1 gram per cubic centimeter. I also assume that the earth is spherical.

Let us first look at the effect of changing the temperature. Let us suppose that the surface temperature of the oceans has increased by an average of 1° Celsius over some long period of time. Lambeck [1980, p. 270] says that the observed changes in the temperature are restricted to the upper 100 or 200 meters of the water. This is primarily intended to apply to changes that occur within a few decades, and may not apply to changes that have occurred over a thousand years. Nonetheless, let us assume that the temperature of the upper 200 meters has increased by 1° and see what happens to the earth's moment of inertia.

We consider a spherical shell of water extending from a radius of 6 378 to 6 378.2 kilometers. The volume of such a shell is 1.0224×10^{17} cubic meters and its mass is 1.0224×10^{20} kilograms. The moment of inertia I of a shell of mass m, with an inner radius r and an outer radius R, is

$$I = (2m/5)(R^5 - r^5)/(R^3 - r^3). \qquad \text{(A.VI.1)}$$

If we divide numerator and denominator by $R - r$, we get

$$I = (2m/5)(R^4 + R^3 r + R^2 r^2 + Rr^3 + r^4)$$
$$\div (R^2 + Rr + r^2).$$

To high accuracy, the numerator of the fraction is $5R^2 r^2$ and the denominator is $3Rr$. Hence

$$I = (2m/3)Rr. \qquad \text{(A.VI.2)}$$

For the case we are considering, I have verified that this is accurate to 8 significant figures. From Equation A.VI.2, we find that

$$I = 2.773 \times 10^{33} \qquad \text{(A.VI.3)}$$

in units of kilograms and meters.

Now suppose that the outer radius R expands while r and m remain constant. The coefficient of expansion of water is a strong function of temperature, and it is not enough to say that the temperature rises by 1°. We must say from what temperature it starts. Most of the moment of inertia of the oceans is contributed by equatorial waters, so it is appropriate to use a moderately high temperature in human terms. I assume that the expansion is from 20° to 21°, and I use the expansion coefficient for pure water rather than salt water.

From a handbook, I find that the specific gravity of water is 0.99823 at 20° and 0.99707 at 25°. If we assume that the change is linear over this range, we find that the volume expands by 0.000 232 8 times the original volume as the temperature changes from 20° to 21°. Let e denote this value, which is the coefficient of volume expansion at the temperature in question:

$$e = 2.328 \times 10^{-4}. \qquad (A.VI.4)$$

Because the inner radius and hence the inner surface area of the shell is assumed not to change, the relative change in volume is the same as the relative change in the thickness of the shell. Thus R changes by 200e = 0.04656 meters. By Equation A.VI.2, the moment of inertia I is proportional to R, so the change δI in I is

$$\delta I = I(200e/R). \qquad (A.VI.5)$$

This gives

$$\delta I = 2.024 \times 10^{25}.$$

Since the moment of inertia I_e of the whole earth is about 8 × 10^{37}, the relative change is 2.5 × 10^{-13}, or

$$\delta I/I_e = 0.000\ 25 \text{ parts in } 10^9 \text{per degree.} \qquad (A.VI.6)$$

This would still be negligible if the change in temperature extended over 10 000 instead of 200 meters.

Now let us look at what happens if there is a change in the amount of glaciation. (The amount of floating ice would have no effect on the moment of inertia if the density of ice were the same as that of water. Since the density of ice is slightly less, there is a small effect, but it is certainly too small to concern us here.) To be specific, suppose we have a mass m of ice at latitude φ and altitude h above sea level. We also suppose that this ice is supported by the solid earth, and that it is not floating.

When this ice melts and flows into the oceans, it affects the moment of inertia of the earth in two ways.

First, by disappearing, it directly decreases the moment of inertia. Second, by appearing in the oceans, it adds to the mass of the oceans and increases their moment of inertia. Its effect on the oceans is to add a thin shell of mass m and radius r. Thus it adds $(2m/3)r^2$ (see Equation A.VI.2) to the moment of inertia. Where it was, its distance from the spin axis was $(r + h) \cos \phi$, so the moment of inertia lost by its melting is $m(r + h)^2 \cos^2 \phi$. The total change δI in the moment of inertia is

$$\delta I = (2/3)mr^2 - m(r + h)^2 \cos^2\phi. \qquad \text{(A.VI.7)}$$

For virtually all ice on the earth, h/r is less than 0.001, so we can neglect h in Equation A.VI.7. Doing so leaves us

$$\delta I = mr^2[(2/3) - \cos^2\phi].$$

This must be summed over all masses of ice. However, for all important bodies of ice, $\cos \phi$ is less than 0.5, and it is sufficient in this exploratory calculation to take

$$\delta I = \tfrac{1}{2}\, mr^2. \qquad \text{(A.VI.8)}$$

The changes in sea level that are occurring at the present time are just at the margin of our present ability to measure, and the results of different workers vary widely. Lambeck [1980, p. 271] takes a rise of 1 millimeter per year as representative of the results. This is a rate of 10 centimeters per century, which amounts to 2 meters in the past 2000 years. In view of the continuity of life in many coastal cities since antiquity, it seems unlikely that the rise in sea level has been more than about this.[†] Let us take a rise of 10 centimeters per century as correct and see where it leads us. Further, let us assume that all this change has come from melting continental ice.

If the sea level rises by 10 centimeters, the mass of water and ice involved is 5.11×10^{16} kilograms, and δI is 1.040×10^{30}. Since the moment of inertia I of the earth is about 8×10^{37}, this gives $\delta I/I = +1.3 \times 10^{-8}$ per century, or $+13$ parts in 10^9 per century. This means that the change in angular velocity is -13 parts in 10^9 per century. That is,

$$y = -13 \qquad \text{(A.VI.9)}$$

if sea level rises by 10 centimeters per century as a result of melting ice.

In deriving Equation A.VI.9, I have neglected the reaction of the solid earth to the melting of the ice and the

----

[†]There is some evidence that sea level has fallen by 2 meters in the past 2000 years. See Section X.5.

consequent expansion of the oceans. Actually, when the ice melts and runs off, it removes a load from the solid earth below, which thereupon rises. This cancels part of the effect of removing the ice. Similarly, when the melt water is added to the oceans, it adds to the load on the ocean floor and makes it sink. This cancels part of the effect of the added water on the moment of inertia. Thus y will be less in magnitude than the value in Equation A.VI.9.

The amount by which y is decreased depends upon the time scale with which the melting occurs. If we are talking about fluctuations of global ice on a time scale of decades, the earth does not have time to react, and Equation A.VI.9 should give a fairly good answer. Jeffreys [1982] in fact thinks that changes in the earth's ice are capable of explaining the decade fluctuations in the earth's rotation, if I understand him correctly.

Jeffreys starts from the measured changes in the earth's rotation and calculates the changes in sea level that would be needed to produce them through the mechanisms of freezing and melting ice. However, he is not able to demonstrate a relation between the changes in rotation and changes in sea level, for lack of sea level data. He can only say that the required changes in sea level seem reasonable.

His calculations are reasonably consistent with Equation A.VI.9. He finds that sea level must fluctuate, on the time scale of a decade, at a rate of about 150 centimeters per century, 15 times the value I used in deriving the equation. This leads to a magnitude of y of 195. In Section I.4, I estimated that the size of y involved in the decade fluctuations is about 350. This is reasonable agreement, considering all the approximations that are involved in both Jeffrey's calculations and mine.

If we are talking about changes on a time scale of millenia, on the other hand, the solid earth does have time to react to the changes in loading, and the size of y in Equation A.VI.9 is too large. However, estimating the reaction of the earth is too complicated a task to undertake here, so I will use Equation A.VI.9 as correct, even though I know it is too large.

APPENDIX VII

SOME EFFECTS OF THE GROWTH OF THE CORE

In this appendix, I calculate the change in the moment
of inertia of the earth that results from the possible
growth of the earth's core at the expense of its mantle. In
doing so, I use Lyttleton's model [Lyttleton, 1982, p. 40]
of the interior of the earth. This model is derived on the
assumption that the earth is not rotating, and that its
figure is a sphere with a radius of 6371 kilometers.

Ths process by which the core can grow is shown in Figure A.VII.1. In part (a) of the figure, we have the exter-
ior surface S_1 of the earth and the interface between the
core and mantle, which is the surface marked C_1 on the inside and M_1 on the outside; we assume with Lyttleton that
the interface has negligible thickness. In part (b) of the
figure, the original interface is shown by the dashed circle
still marked C_1, but we assume that the core has grown from
C_1 to the surface marked C_2. Let us assume that this growth
takes place in negligible time.

Let ρ_c denote the density of the core material just
below the interface and let ρ_m denote the density of the
mantle material just above it. Since ρ_c is greater than
ρ_m, the mantle loses more volume than the core gains. Hence
the lower surface of the mantle (part (b)) must have re-
treated to a surface M_2 that is no longer in contact with
the core at C_2. That is, immediately after the core grows
from C_1 to C_2, there is an empty shell between C_2 and M_2.
However, this state of affairs cannot persist, so there must
be a stage of adjustment.

Because of the load imposed on it by the rest of the
mantle, the material now at the bottom of the mantle must
fail structurally and move down into the once-empty shell.
This leaves an empty space above the bottom of the mantle,
which is filled by material from farther up, and so on.
Thus we have a wave spreading out through the mantle, not
necessarily homogeneously, until it reaches the surface
(part (c)). The net result of the whole process is that the
surface shrinks from S_1 to S_2, and the mass in the shell
between S_1 and S_2 is transferred to the shell between C_1 and
C_2, increasing its density from ρ_m to ρ_c. C_2 becomes the
new interface between the core and mantle.

Now let us work out the change in the moment of inertia
that results if we add 1 meter to the radius of the core.
Let r_c be the initial radius of the core, measured in meters. The area $\underline{A}$ of the surface C_1 is

$$\underline{A} = 4\pi \underline{r}_c{}^2, \hspace{3cm} (A.VII.1)$$

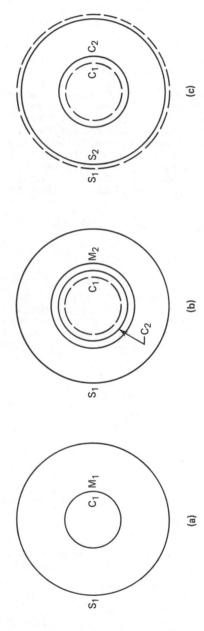

(a)

(b)

(c)

Figure A.VII.1. The growth of the core. At some time (part a), the interface between the core and mantle is at the surface marked both C_1 and M_1. The core then grows to the surface marked C_2, depleting all the mantle between M_1 and M_2 as it does so (part b). By collapse of material above it, the space between C_2 and M_2 is filled in (part c), ultimately at the expense of the surface. The final result is that the mass originally between S_1 and S_2 is added to the shell between C_1 and C_2, increasing the density of the latter and decreasing the moment of inertia of the earth. C_2 becomes the new interface.

and this is also the volume of a layer 1 meter thick. The
density of this layer changes from ρ_m to ρ_c, and the change
$\delta\rho$ is $\rho_c - \rho_m$. Hence the change $\delta\underline{m}$ of the mass that lies in
the shell between C_1 and C_2 is

$$\delta\underline{m} = \underline{A}\delta\rho. \qquad (A.VII.2)$$

This is also the mass originally in the shell between the
surfaces S_1 and S_2.

Before the transition, the moment of inertia of the
mass $\delta\underline{m}$ was (see Equation A.VI.2) $(2/3)\delta\underline{m}r_{\underline{e}}^2$, in which $r_{\underline{e}}$
is the radius of the surface S_1. After the transition, the
moment of inertia equals $(2/3)\delta\underline{m}r_{\underline{c}}^2$. Hence the change $\delta\underline{I}$
in the moment of inertia is

$$\delta\underline{I} = (2/3)\delta\underline{m}(r_{\underline{c}}^2 - r_{\underline{e}}^2). \qquad (A.VII.3)$$

Lyttleton [p. 40] adopts the following values:

$$r_{\underline{c}} = 3473 \text{ kilometers}, \qquad \rho_{\underline{c}} = 9430 \text{ kilograms/meter}^3,$$
$$\qquad\qquad\qquad\qquad\qquad\qquad\qquad\qquad\qquad\qquad (A.VII.4)$$
$$r_{\underline{e}} = 6371 \text{ kilometers}, \qquad \rho_{\underline{m}} = 5680 \text{ kilograms/meter}^3.$$

If we substitute these values into Equations A.VII.1 through
A.VII.3, we get

$$\delta\underline{I} = -1.081 \times 10^{31} \quad \text{kilogram-meter}^2, \qquad (A.VII.5)$$

for a change of 1 meter in the radius of the core.

Now let us estimate the possible rate of growth of $r_{\underline{c}}$.
Lyttleton [p.100] estimates that the core grew from nothing
to a radius of 2042 kilometers in a time that is negligible
in the history of the earth. He estimates [p. 127] that
this happened at a time that is between 2.5 and 4.0 billion
years ago. Hence the rate $\dot{r}_{\underline{c}}$ is given by

$$0.0358 < \dot{r}_{\underline{c}} < 0.0572, \qquad (A.VII.6)$$

in meters per century.

If we further use the fact that the present moment of
inertia $\underline{I}$ of the earth is about 8×10^{37} kilogram-meter^2,
and calculate the rate of change $\underline{I}$ of the earth, we find

$$4.83 < (-\dot{\underline{I}}/\underline{I}) < 7.73 \qquad (A.VII.7)$$

parts in 10^9 per century.

In calculating Equation A.VII.7, I have assumed that the density ρ_m at the bottom of the mantle does not change as a consequence of all these events. Actually, since some mass is lost to the mantle, the loading at the bottom of the mantle decreases and so does the density ρ_m. However, the relative change in density is of the order of δm (Equation A.VII.2) divided by the mass of the earth, so the effect of changing the density ρ_m is of second order.

REFERENCES

(Parts of a name like van, ibn, and al (the latter two
in Islamic names) are considered as particles and do not
appear as the initial element in most alphabetic listings.
Indeed, in a work like a biographical dictionary, there are
so many names starting with al, for example, that using it
as the initial element would be impractical. In this list
of references, however, there are not many entries beginning
with these particles. Since most people tend to think of
the particle as the first element in the name, I have usually treated it that way in this list. Thus, for example,
al-Battani appears under A rather than B.)

Aaboe, A., Observation and theory in Babylonian astronomy,
    Centaurus, 24, pp. 14-35, 1980.

al-Battani, Abu Abd Allah Muhammad ibn Jabir ibn Sinan al-
    Raqqi al-Harrani al-Sabi, De Scientia Stellarum Liber,
    ca. 925. There is a translation into Latin (with the
    author's name spelled Albategnius) by Plato of Tivoli,
    Typis Haeredis Victorii Benatii, Bologna, 1645. There
    is a modern edition of the Arabic text, with a trans-
    lation into Latin, by C. A. Nallino, Pubblicazione Reale
    Osservatorio di Brera, Milan, in 3 volumes, 1899-1907.
    (See Hartner [1970] for a discussion of this edition.)

al-Biruni, Abu-Rayhan Muhammad bin Ahmad, Kitab Tahdid Niha-
    yat al-Amakin Litashih Masafat al-Masakin, 1025. There
    is a translation into English, under the title The
    Determination of the Coordinates of Positions for the
    Correction of Distances between Cities, by Jamil Ali,
    American University of Beirut, Beirut, Lebanon, 1967.
    Page references are to this translation.

Allen, C. W., Astrophysical Quantities, Second Edition, Ath-
    lone Press, London, 1962.

American Ephemeris and Nautical Almanac, U.S. Government
    Printing Office, Washington, published annually. This
    is now called The Astronomical Almanac.

Andrieu, J., César; Guerre d'Alexandrie, an edition and
    parallel translation into French, with critical notes,
    of Hirtius [ca. -43], Société d'Edition "Les Belles
    Lettres", Paris, 1954.

Armitage, A., The World of Copernicus, Mentor Books, New
    York, 1951. This was originally published in 1947 by
    Henry Schuman, Inc., under the title Sun, Stand Thou
    Still.

Ashbrook, J., Review of Medieval Chronicles and the Rotation
    of the Earth,† Sky and Telescope, 44, p. 85, August
    1972.

---

†This is the work cited here as Newton [1972].

Beaver, D. deB., Bernard Walther: Innovator in astronomical observation, Journal for the History of Astronomy, i, pp. 39-43, 1970.

Becvar, A., Atlas of the Heavens, Catalogus 1950.0, Czechoslovak Academy of Sciences, Prague, or Sky Publishing Corporation, Cambridge, Massachusetts, 1964.

Brahe, Tycho, Astronomiae Instauratae Mechanica, Philip de Ohr, Wandsbeck, Prussia, 1598. There is a reprint by Levinus Hulsius, Nürnberg, 1602; I used a copy of this reprint in preparing the present work. There is another reprint edited by J. L. E. Dreyer in Dreyer [1926, volume V, pp. 1-162]. There is a translation of the main parts of the work into English, with the title given as Tycho Brahe's Description of His Instruments and Scientific Work, by H. Raeder, E. Strömgren, and B. Strömgren, I Kommission Hos Ejnar Munksgaard, København, 1946. (The translators omit the dedication and some of Tycho's poems, but they include everything that is directly relevant to astronomy.)

Britton, J. P., On the quality of solar and lunar observations and parameters in Ptolemy's Almagest, a dissertation submitted to Yale University, 1967.

Britton, J. P., Ptolemy's determination of the obliquity of the ecliptic, Centaurus, 14, pp. 29-41, 1969.

Brouwer, D., A study of the changes in the rate of rotation of the earth, Astronomical Journal, 57, pp. 125-146, 1952.

Brown, E. W., with the assistance of H. B. Hedrick, Chief Computer, Tables of the Motion of the Moon, in 3 volumes, Yale University Press, New Haven, Connecticut, 1919. (The reader should also note Complement to the Tables of the Moon, Containing the Remainder Terms for the Century 1800-1900, and Errata in the Tables, with the same authors and publisher, 1926).

Caesar, C. J., De Bello Civili, ca. -43. There is an edition in C. Julii Caesaris, Quae Extant,† Carey, Johnson, Warner, et al., Philadelphia, 1813. (It is probable that this work was not actually published until after Caesar's death in -43).

Censorinus, Liber de Die Natali, 238. There is an edition, with a parallel translation into French, by Desiré Nisard in Celse, Vitruve, Censorin, et Frontin, Dubochet et Cie., Paris, 1846.

Clemence, G. M., First-order theory of Mars, Astronomical Papers Prepared for the Use of the American Ephemeris and Nautical Almanac, XI, Part 2, U.S. Government Printing Office, Washington, 1949.

---

†It seems to me that the word Opera should occur somewhere in this title, but it does not.

Clemence, G. M., Theory of Mars - completion, Astronomical Papers Prepared for the Use of the American Ephemeris and Nautical Almanac, XVI, Part 2, U.S. Government Printing Office, Washington, 1961.

Cohen, A. P. and Newton, R. R., Solar eclipses recorded in China during the Tarng dynasty, Monumenta Serica, 35, in press.

Copernicus, Nicolaus, De Revolutionibus Orbium Caelestium Libri Sex, J. Petrejus, Nürnberg, 1543. There is an edition by Fritz Kubach, Verlag von R. Oldenbourg, Munich, in two volumes: volume 1, a facsimile of Copernicus's holograph, 1944; volume 2, a critical edition of the text, 1949. There is a translation into English by C. G. Wallis, Great Books of the Western World, volume 16, Encyclopaedia Britannica, Inc., Chicago, 1952.

Curtz, Albert, Historia Coelestis, Augsburg (publisher not stated), 1666. (The reader may find this work cited under the name of Curtz, Barethis, Barrettus, Tycho, or Brahe. See the discussion in Section III.1.)

Delambre, J. B. J., Histoire de l'Astronomie du Moyen Âge, Chez Mme. Veuve Courcier, Paris, 1819.

Dreyer, J. L. E., Tycho Brahe, Adam and Charles Black, Edinburgh, 1890.

Dreyer, J. L. E., On the original form of the Alfonsine Tables, Monthly Notices of the Royal Astronomical Society, 80, pp. 243-262, 1920.

Dreyer, J. L. E., Tychonis Brahe Dani Opera Omnia, in 15 volumes, Libraria Gyldendaliana, København, 1926. The various volumes appeared over a span of many years, but I use 1926 in citation because it is the year when the last volume of Tycho's observations appeared. Volume 14 of the series, edited by E. Nystrom, appeared in 1928, and volume 15, edited by Dreyer and H. Raeder, appeared in 1929. These are not relevant to the present work and I have not cited them. The interested reader may wish to consult "Addenda to Tychonis Brahe Opera Omnia tomus XIV, by J. R. Christianson, in Centaurus, 16, pp. 231-249, 1972.

Dubs, H. H., Solar eclipses during the Former Han period, Osiris, 5, pp. 499-522, 1938.

Eckert, W. J., Brouwer, D., and Clemence, G. M., Coordinates of the five outer planets 1653 - 2060, Astronomical Papers Prepared for the Use of the American Ephemeris and Nautical Alamanac, XII, U.S. Government Printing Office, Washington, 1951.

Eckert, W. J., Jones, Rebecca, and Clark, H. K., Construction of the lunar ephemeris, in An Improved Lunar Ephemeris, 1952 - 1959, issued as a Joint Supplement to the American Ephemeris and Nautical Almanac and the Nautical Almanac and Astronomical Ephemeris, U.S. Government Printing Office, Washington, 1954.

Epping, J. and Strassmaier, J. N., Neue babylonische Planeten-Tafeln, II Teil, Zeitschrift für Assyriologie and Verwandte Gebiete, 6, pp. 89-102, 1891a.

Epping, J. and Strassmaier, J. N., Neue babylonische Planeten-Tafeln, III Teil, Zeitschrift für Assyriologie und Verwandte Gebiete, 6, pp. 217-228, 1891b.

Epping, J. and Strassmaier, J. N., Babylonische Mondbeobach- tungen aus den Jahren 38 und 79 der Seleuciden-Aera, Zeitschrift für Assyriologie und Verwandte Gebiete, 7, pp. 220-254, 1892.

Explanatory Supplement to The Astronomical Ephemeris and The American Ephemeris and Nautical Almanac, H. M. Station- ery Office, London, fourth printing with amendments, 1977.

Fotheringham, J. K. (assisted by Gertrude Longbottom), The secular acceleration of the moon's mean motion as deter- mined from the occultations in the Almagest, Monthly Notices of the Royal Astronomical Society, 75, pp. 377- 394, 1915.

Fotheringham, J. K., A solution of ancient eclipses of the sun, Monthly Notices of the Royal Astronomical Society, 81, pp. 104-126, 1920.

Fotheringham, J. K., Two Babylonian eclipses, Monthly Noti- ces of the Royal Astronomical Society, 95, supplementary number, pp. 719-723, 1935.

Goldstein, B. R., The astronomical tables of Levi ben Ger- son, Transactions of the Connecticut Academy of Arts and Sciences, 45, 1974.

Goldstein, B. R., Levi ben Gerson's preliminary lunar model, Centaurus, 18, pp. 275-288, 1974a.

Goldstein, B. R., Levi ben Gerson's analysis of precession, Journal for the History of Astronomy, vi, pp. 31-41, 1975.

Goldstein, S. J., Jr., Deceleration of the earth's rotation from old solar tables, Celestial Mechanics, 27, pp. 53- 63, 1982.

Hartner, W., article on al-Battani in the Dictionary of Scientific Biography, volume I, Charles Scribner's Sons, New York, 1970.

Haskins, C. H., Studies in the History of Medieval Science, Harvard University Press, Cambridge, Massachusetts, 1924.

Hellman, C. Doris, article on Tycho Brahe in the Dictionary of Scientific Biography, volume II, Charles Scribner's Sons, New York, 1970.

Henriksen, S. W., The hydrostatic flattening of the earth, Annals of the International Geophysical Year, volume XII, part 1, pp. 197-198, Pergamon Press, New York, 1960.

Hirtius, A., Commentariorum de Bello Alexandrino Liber, ca. -43.† There is an edition in C. Julii Caesaris, Quae Extant, Carey, Johnson, Warner, et al., Philadelphia, 1813. See Andrieu [1954] and Caesar [ca. -43].

ibn Yunis, Abul-Hasan Ali ibn Abd al-Rahman ibn Ahmad ibn Yunis al-Sadafi, Az-Zij al-Kabir al-Hakimi, 1008. There is an edition of part of this work, with a parallel translation into French, by J. J. A. Caussin de Perceval, under the title Le Livre de la Grande Table Haké-mite, with the author's name spelled as Ebn Iounis, Imprimerie de la République, Paris, 1804. Page cita-tions are to this edition.

Jeffreys, H., Tidal friction; the core; mountain and conti-nent formation, Geophysical Journal of the Royal Astro-nomical Society, 71, pp. 555-566, 1982.

Kennedy, E. S., A survey of Islamic tables, Transactions of the American Philosophical Society, 46, no. 2, pp. 123-175, 1956.

Kennedy, E. S., article on al-Biruni in the Dictionary of Scientific Biography, volume II, Charles Scribner's Sons, New York, 1970.

King, D. A., article on ibn Yunus (note spelling) in the Dictionary of Scientific Biography, volume XIV, Charles Scribner's Sons, New York, 1976.

Knobel, E. B., Notes on a Persian MS. of Ulugh Beigh's cata-logue of stars belonging to the Royal Astronomical Soci-ety, Monthly Notices of the Royal Astronomical Society, 39, pp. 337-363, 1879.

Kovacheva, Mary, Archaeomagnetic investigations of geomag-netic secular variations, Philosophical Transactions of the Royal Society of London, A, 306, pp. 79-86, 1982.

---

†Aulus Hirtius was a prominent Roman of the time, and a close friend and partisan of Caesar. His book about the Alexandrian campaign is often bound with the works of Caesar. Some authorities believe that the true author was not Hirtius but another friend of Caesar named Gaius Oppius.

Kremer, R. L., Bernard Walther's astronomical observations, Journal for the History of Astronomy, xi, pp. 174-191, 1980.

Kremer, R. L., The use of Bernard Walther's astronomical observations: theory and observation in early modern astronomy, Journal for the History of Astronomy, xii, pp. 124-132, 1981.

Kugler, F. X., Sternkunde und Sterndienst in Babel, I Teil, Aschendorffsche Verlagsbuchhandlung, Münster, Westfalen, 1907.

Kugler, F. X., Sternkunde und Sterndienst in Babel, Ergänzungen zum Ersten und Zweiten Buch, II Teil, Aschendorffsche Verlagsbuchhandlung, Münster, Westfalen, 1914.

Kugler, F. X., Sternkunde und Sterndienst in Babel, II Teil, 2 Heft, Aschendorffsche Verlagsbuchhandlung, Münster, Westfalen, 1924.

Lambeck, K., The Earth's Variable Rotation: Geophysical Causes and Consequences, Cambridge University Press, Cambridge, 1980.

Lyttleton, R. A., The Earth and Its Mountains, John Wiley & Sons, New York, 1982.

Markowitz, W., Sudden changes in rotational acceleration of the earth and secular motion of the pole, in Earthquake Fields and the Rotation of the Earth, L. Mansinha, D. E. Smylie, and A. E. Beck, editors, pp. 69-81, D. Reidel Publishing Co., Dordrecht, Holland, 1970.

Martin, C. F., A study of the rate of rotation of the earth from occultations of stars by the moon, 1627-1860, a dissertation presented to Yale University, 1969.

Muller, P. M., An analysis of the ancient astronomical observations with the implications for geophysics and cosmology, a dissertation presented to the School of Physics, University of Newcastle, Newcastle-upon-Tyne, 1975.

Muller, P. M. and Stephenson, F. R., The accelerations of the earth and moon from early astronomical observations, in Growth Rhythms and the History of the Earth's Rotation, G. D. Rosenberg and S. K. Runcorn, editors, pp. 459-533, John Wiley and Sons, New York, 1975.

Munk, W. H. and MacDonald, G. J. F., The Rotation of the Earth, Cambridge University Press, Cambridge, 1960.

National Oceanic and Atmospheric Administration, Monthly Climatic Data for the World, sponsored by the World Meteorological Organization, published monthly by the National Oceanic and Atmospheric Administration at Asheville, North Carolina, 1976. (I use only the issues for 1976 in this work.)

Neugebauer, O., Astronomical Cuneiform Texts, in 3 volumes, Lund Humphries, London, 1955.

Neugebauer, O., The Exact Sciences in Antiquity, Second Edition, Brown University Press, Providence, Rhode Island, 1957.

Neugebauer, O., The Astronomical Tables of al-Khwarizmi, in Historisk-filosofiske Skrifter, Kongelige Danske Videnskabernes Selskab, 4, no. 2, 1962.

Neugebauer, O., A History of Ancient Mathematical Astronomy, Springer-Verlag, New York, 1975.

Neugebauer, P. V. and Weidner, E. F., Ein astronomischer Beobachtungstext aus dem 37. Jahre Nebukadnezars II. (-567/66), Berichte über die Verhandlungen der Königlichen Sachsischen Akademie der Wissenschaften zu Leipzig, Philologie-Historie Klasse, 67, Heft 2, pp. 29-89, 1915.

Newcomb, S., Researches on the Motion of the Moon, in Washington Observations, U.S. Naval Observatory, Washington, 1875.

Newcomb, S., Tables of the motion of the earth on its axis and around the sun, Astronomical Papers Prepared for the Use of the American Ephemeris and Nautical Almanac, VI, Part 1, U.S. Government Printing Office, Washington, 1895.

Newcomb, S., Tables of Mercury, Astronomical Papers Prepared for the Use of the American Ephemeris and Nautical Almanac, VI, Part 2, U.S. Government Printing Office, Washington, 1895a.

Newcomb, S., Tables of Venus, Astronomical Papers Prepared for the Use of the American Ephemeris and Nautical Almanac, VI, Part 3, U.S. Government Printing Office, Washington, 1895b.

Newhall, X. X., Williams, J. G., and Standish, E. M., The construction of the DE102 ephemeris, paper intended for the Astronomical Journal, in preparation.

Newton, R. R., Ancient Astronomical Observations and the Accelerations of the Earth and Moon, Johns Hopkins University Press, Baltimore and London, 1970.

Newton, R. R., Medieval Chronicles and the Rotation of the Earth, Johns Hopkins University Press, Baltimore and London, 1972.

Newton, R. R., The earth's acceleration as deduced from al-Biruni's solar data, Memoirs of the Royal Astronomical Society, 76, pp. 99-128, 1972a.

Newton, R.. R., Astronomical evidence concerning non-
gravitational forces in the earth-moon system, Astrophysics and Space Science, 16, pp. 179-200, 1972b.

Newton, R. R., The historical acceleration of the earth,
Geophysical Surveys, 1, pp. 123-145, 1973.

Newton, R. R., Ancient Planetary Observations and the Validity of Ephemeris Time, Johns Hopkins University Press,
Baltimore and London, 1976.

Newton, R. R., The Crime of Claudius Ptolemy, Johns Hopkins
University Press, Baltimore and London, 1977.

Newton, R. R., The Moon's Acceleration and Its Physical Origins, volume 1, Johns Hopkins University Press, Baltimore and London, 1979.

Newton, R. R., An analysis of the solar observations of
Regiomontanus and Walther, Quarterly Journal of the
Royal Astronomical Society, 23, pp. 67-93, 1982.

Newton, R. R., The Origins of Ptolemy's Astronomical Parameters, Technical Publication no. 4, Center for Archaeo-
astronomy, University of Maryland, College Park, Maryland, 1983.

North, J. D., article on Hevelius in the Dictionary of Scientific Biography, volume VI, Charles Scribner's Sons,
New York, 1972.

Oppolzer, T. R. von, Canon der Finsternisse, Kaiserlich-
Königlichen Hof- und Staatsdruckerei, Wien, 1887. There
is a reprint, with the introduction translated into English by O. Gingerich, by Dover Publications, New York,
1962.

Pannekoek, A., A History of Astronomy, Interscience Publish-
ers, New York, 1961.

Parker, R. A. and Dubberstein, W. H., Babylonian Chronology,
626 B.C.-A.D. 75, Brown University Press, Providence,
Rhode Island, 1956.

Pauly-Wissowa [1894]. I use this as a convenient citation
for all volumes of Paulys Real-Encyclopadie der Class-
ischen Altertumswissenschaft, J. B. Metzler, Stuttgart,
1894 and later years. (This was started by August Pauly
in about 1839, continued by Georg Wissowa in 1894 and
later years, and continued by still others after
Wissowa.)

Pederson, O., A Survey of the Almagest, Odense University
Press, Odense, Denmark, 1974.

Plutarch, Life of (Julius) Caesar, ca. 100. There is a translation into English in the Harvard Classics, P. F. Collier and Son Company, New York, 1909.

Ptolemy, C., 'E Mathematike Syntaxis, ca. 142. There is an edition by J. L. Heiberg in C. Ptolemaei Opera Quae Exstant Omnia, B. G. Teubner, Leipzig, 1898. There is a translation of this edition into German by K. Manitius, B. G. Teubner, Leipzig, 1913. There is also a translation of this edition into English by R. C. Taliaferro in volume 16 of Great Books of the Western World, Encyclopaedia Britannica, Inc., Chicago, 1952.

Rawlins, D., An investigation of the ancient star catalog, Publications of the Astronomical Society of the Pacific, 94, pp. 359-373, 1982.

Rico, Manuel (Rico) y Sinobas, Libros del Saber de Astronomia del Rey Don Alfonso X. de Castilla, in five volumes, Tipografia de Don Eusebio Aguado, Madrid, 1867.†

Rosen, E., article on Copernicus in the Dictionary of Scientific Biography, volume III, Charles Scribner's Sons, New York, 1971.

Rosen, E., article on Regiomontanus in the Dictionary of Scientific Biography, volume XI, Charles Scribner's Sons, New York, 1975.

Sachs, A., A classification of the Babylonian astronomical tablets of the Seleucid period, Journal of Cuneiform Studies, 2, pp. 271-290, 1948.

Schaumberger, P. J., Drei babylonische Planetentafeln der Seleukidenzeit, Orientalia, 2, pp. 97-116, 1933.

Schmeidler, F., Johannes Regiomontanus, Opera Collectanea, Otto Zeller Verlag, Osnabruck, 1972. (This reference is often cited or catalogued as if Regiomontanus were the author. See the discussion in Section III.1)

Simeon of Durham (Symeonis monachi Dunelmensis), Historia Regum Anglorum et Dacorum, ca. 1129. There is an edition with notes by Thomas Arnold in Rerum Britannicarum Medii Aevi Scriptores, number 75, volume 2, Longmans and Company, London, 1885.

Smart, W. M., Text-book on Spherical Astronomy, Fifth Edition, Cambridge University Press, Cambridge, 1962. (The first edition was in 1931.)

Smith, P. J., The intensity of the ancient geomagnetic field: a review and analysis, Geophysical Journal of the Royal Astronomical Society, 12, pp. 321-362, 1967.

---

†This work is often catalogued under the name of Alfonso.

Spencer Jones, H., The rotation of the earth, and the secular accelerations of the sun, moon, and planets, Monthly Notices of the Royal Astronomical Society, 99, pp. 541-558, 1939.

Strabo, Geography, ca. 0. There is an edition, with a parallel translation into English, by H. L. Jones, in 8 volumes, G. P. Putnam's Sons, New York, 1932.

Syntaxis = Ptolemy [ca. 142].

Thoren, V. E., New light on Tycho's instruments, Journal for the History of Astronomy, iv, pp. 25-45, 1973.

Times Atlas of the World, Mid-Century Edition, in 5 volumes, The Times Office, London, 1955.

Toomer, G. J., article on al-Khwarizmi in the Dictionary of Scientific Biography, volume VII, Charles Scribner's Sons, New York, 1973.

Tupman, G. L., A comparison of Tycho Brahe's meridian observations of the sun with Leverrier's solar tables, The Observatory, XXIII, pp. 132-135, continued on pp. 165-171, 1900.

van der Waerden, B. L., The Birth of Astronomy, which is volume 2 of Science Awakening, Oxford University Press, New York, 1974.

Van Flandern, T. C., Improved elements for the sun and moon, paper intended for the Astronomical Journal, in preparation.

Webster's Geographical Dictionary, G. & C. Merriam Co., Springfield, Massachusetts, 1949.

Wesley, W. G., The accuracy of Tycho Brahe's instruments, Journal for the History of Astronomy, ix, pp. 42-53, 1978.

Williams, J. G., Sinclair, W. S., and Yoder, C. F., Tidal acceleration of the moon, Geophysical Research Letters, 5, pp. 943-946, 1978.

Wylie, A., Chinese Researches, (publisher not stated), Shanghai, 1897. Reprinted by Ch'eng Wen Publishing Co., Taipei, 1966.

Yabuuchi, K., Zui Tō Rekihōshi no Kenkyū (Researches on the History of Calendrical Methods of the Swei and Tarng Periods), Sanseido, Tokyo, 1944.

INDEX